Place Names in Africa

Liora Bigon
Editor

Place Names in Africa

Colonial Urban Legacies, Entangled Histories

Springer

Editor
Liora Bigon
The Institute of Western Cultures
The Hebrew University of Jerusalem
Jerusalem, Israel

Holon Institute of Technology
Holon, Israel

ISBN 978-3-319-32484-5 ISBN 978-3-319-32485-2 (eBook)
DOI 10.1007/978-3-319-32485-2

Library of Congress Control Number: 2016940553

Printed on acid-free paper

This Springer imprint is published by Springer Nature
The registered company is Springer International Publishing AG Switzerland

Foreword

Over the past two decades, there have been numerous efforts to reorient the field of place-name studies toward a critical account of the social and political struggles over toponymic inscription by situating such practices within the broader politics of space and place. This edited collection provides a welcome addition to the emerging body of scholarship on critical toponymies while also drawing important linkages with urban planning history and postcolonial studies. As Liora Bigon rightly points out in the introductory chapter, critical toponymic scholarship has drawn considerable attention to the contested politics of place naming in Europe and North America, yet far less consideration has been given to other world regions such as Africa. This volume, therefore, fills a significant gap in the literature by bringing African toponymies to the foreground through a series of spatial histories that explore the cultural entanglements of indigenous, colonial, and postcolonial regimes of toponymic inscription.

The book's emphasis on the "colonial legacies" and "entangled histories" of place naming acknowledges the complexity of the toponymic tapestry in African cities. This is nicely illustrated by a combined focus on both the long-term historical structures shaping toponymic landscapes over the course of centuries (*longue durée*) as well as the history of events (*l'histoire événementielle*) that have transformed the symbolic infrastructure of urban spaces. Pivotal "events", such as revolutionary changes in political regime, often bring about radical transformations to a city's toponymy, as various contributions to this book show. Yet the present collection also highlights the ways in which multiple layers of meaning and memory – both official and vernacular – often coexist simultaneously in the very same cityscape.

Taken together, the contributions to this book offer a wide ranging set of case studies from across the African continent that enrich our understanding of the spatial narratives and geographical imaginaries that have been etched into urban spaces by colonial regimes and their postcolonial successors. The result is an interdisciplinary collection of essays that draws upon the insights of urban history, sociolinguistics, political science, and cultural geography, among other fields. Given this diverse cast of authors, it is to be expected that each will approach the subject matter in a somewhat different fashion. Some authors are interested in tracing the genealogy of individual toponyms while other contributors draw upon semiotics and postcolonial theory to unpack the symbolic power of entire regimes of toponymic inscription. There is also a diverse array of methodological approaches – from archival research to ethnographic fieldwork – that informs the studies in this volume. A number of contributors have adopted a "top-down" approach highlighting the administrative procedures that shape the official naming of places, whereas other chapters underscore the importance of "bottom-up" processes of vernacular speech that constitute the toponymic spaces of everyday life.

This collection also broadens the focus of critical place-name studies beyond "specific" toponyms by also considering a number of "generic" naming practices that played a significant role in the social ordering of urban space under colonialism in Africa. For example, Luce Beeckmans's account of the establishment of so-called neutral zones – which were anything but "neutral" – nicely demonstrates how the use of euphemistic, generic place names was central to the legitimization of the white supremacist doctrines of racial segregation that underpinned colonial planning practices. Similarly, Liora Bigon and Yossi Katz's chapter on the transplantation of "garden city" discourses from Europe to the Global South is equally instructive in showing how generic toponymies were employed to literally *naturalize* the geographical divisions between colonizer and colonized.

Another important contribution that this edited collection makes to critical toponymic scholarship is its emphasis on the relationality of place naming. As various authors in this volume contend, the naming of places in one location cannot be fully grasped in isolation. This is true on multiple levels. For example, a toponym can link multiple places together through long-distance trade (as in the case of Walter Brown's account of Bagamoyo in East Africa). Additionally, colonial authorities in Africa often adopted similar place-naming practices in different colonial cities, which makes relational critiques of toponymic "regimes" all the more salient. And if one needed any further evidence that place naming should be understood in relational terms, Wale Adebanwi's provocative account of the retaliatory renaming of streets in New York and Lagos provides a compelling case that place naming can be far more than a local affair alone. Put simply, toponyms must be understood not as isolated linguistic objects but rather as inscriptions which are enmeshed in socio-spatial processes that are best grasped relationally.

Yet beyond its theoretical and methodological contributions, it's worth reiterating that one of the chief advances that this book makes is to call into question the Eurocentrism that pervades critical toponymic scholarship. To be sure, there have been a number of critical studies of place naming in the African context, yet, up until this point, such studies have been largely scattered across different scholarly journals that are tailored to specific disciplinary audiences. To my knowledge, this collection is one of the first books to be published that brings together into a single volume the interdisciplinary literature on place naming in Africa. In doing so, the present book broadens the geographical scope of critical place-name studies considerably and will hopefully make such works more accessible to scholars and the broader public alike.

Victoria, BC
August 2015

Reuben Rose-Redwood

Acknowledgements

This collection is first and foremost the fruit of the team effort of its seventeen participants, bridging between challenging multilateral channels of communication, a variety of languages, a variegated use of primary and secondary sources, and work in situ. I would like to warmly thank all the participants.

Special thanks are due to Professor Ruth Fine and the Institute of Western Cultures at the Hebrew University of Jerusalem for their encouragement of this research initiative and their provision of the necessary financial support during all the project's stages. The Society of Architectural Historians of Great Britain (SAHGB) also provided some financial assistance. Further administrative and contextual support was provided by Professor Noam Shoval and Professor Ruth Kark from the Department of Geography of the Hebrew University.

This collection and the motives behind its creation were primarily designed to correspond with some thematic, geographic, methodological, and historiographic gaps in the field of toponymic studies today (see introductory chapter). It is a result of a few thoughtful years regarding place names and Africa, which yielded, inter alia, a series of conference papers that were authored by the Editor or co-authored especially with Professor Ambe Njoh from the University of South Florida and Professor Yossi Katz from Bar Ilan University. These papers were presented in a variety of platforms, e.g., the European Association for Urban History (EAUH): a specialist session entitled: "What's in a name? How do we label peripheral places", Prague, 2012, and a main session entitled "Ubiquitous yet unique: green spaces in cities from the eighteenth century to the present", Lisbon, 2014; the International Planning History Society (IPHS) conference "Colonial and postcolonial urban planning in Africa", Lisbon, 2013; and the 54 Israeli Geographical Society's Annual Conference: a session entitled "Toponymic practices in global view", Tel Aviv, 2013 – I would like to thank the valuable and enlightening comments received from colleagues participants.

Many other people have contributed to the production of this collection, more or less directly, with inspirational conversations, interchange of ideas and additional professional support and technical assistance. A special gratitude is recoded to: Ester Adamit, Ademide Adelusi-Adeluyi, Salih Akin, Victor Azarya, Yoram Bar-Gal,

Meron Benvenisti, Dorothee Brantz, Kara Brown, Stephan Bühnen, Fatoumata Cissé, Peter Clark, Alan Clayman, Amer Dahanshe, Louis Dioh, Nigel Gibson, Hervé Guillorel, Richard Harries, Sonia Hirt, Ya'akov Maoz, Helen Meller, Elizabeth Mimiafou, Tia Mysoa, Deborah Pellow, Sean Purdy, RCC Company Nigeria, Reuben Rose-Redwood, Uvi Rubinstein, Awa Seck, Claire Simonneau, and Charlotte Vorms.

Further acknowledgments, as relevant to each contribution – whether regarding republishing permission in a few cases, other copyright issues as to the images, cartographic design, etc. – are explicitly indicated in the appropriate place within the chapters. I am also indebted to the critical encouragement made through Springer, and to the team effort of everyone at this Press, particularly the Publishing Editor in geography Margaret Deignan and her assistant, Timo Bazuin.

Contents

Contributors

Wale Adebanwi is an associate professor in African American and African Studies, University of California, Davis, United States. He holds a PhD in Social Anthropology from the University of Cambridge, UK, and another PhD in Political Science from the University of Ibadan, Nigeria. He has taught political science at the University of Ibadan and he is a Visiting Professor at Rhodes University, Grahamstown, South Africa. He is the author of *Yoruba Elites and Ethnic Politics in Nigeria: Obafemi Awolowo and Corporate Agency* (2014) and *Authority Stealing: Anti-corruption War and Democratic Politics in Post-military Nigeria* (2012). He has published articles on cities, territoriality, and spatial and urban politics.

Luce Beeckmans was graduated as a civil engineer/architect in 2005 from the Department of Architecture and Urban Planning, Ghent University, Belgium. She then worked as an architect and urban planner with the office of Stéphane Beel Architects. In 2013, she obtained an Arts PhD at Groningen University, the Netherlands. In her doctoral dissertation "Making the African City", she studied urban (planning) history in Africa from a comparative and interdisciplinary perspective. Her dissertation was awarded the Jan van Gelder Prize for best art history publication of the year in 2013. Since then she has been a full-time post-doctoral researcher at the Department of Architecture and Urban Planning of Ghent University.

Michel Ben Arrous is an independent scholar. He has lectured and published extensively on the epistemology of territory, media and conflict, and the history of geographic ideas and fantasies. North-Africa born, an architect by training (Ecole Polytechnique Fédérale de Lausanne, EPFL), a doctor in geography (PhD Université de Rouen) and a former journalist in Southern and West Africa, he also spent many years as research coordinator with CODESRIA (Council for the development of social science research in Africa) in Dakar, Senegal. His current research interests focus on African and Jewish practices and concepts of space, and the comparative history of geographic representations. He is also the editor of the literary blog nobelstudio.org.

Liora Bigon (PhD in Architecture, Manchester University, UK) is a Research Fellow at the Institute of Western Cultures, the Hebrew University of Jerusalem and Holon Institute of Technology, Israel. A specialist in colonial urban history, architecture, and planning cultures in sub-Saharan Africa, she has taught in several universities and academic institutions and published widely in these fields, including articles, encyclopedic entries, and books (*A History of Urban Planning in Two West African Colonial Capitals*, Mellen, 2009; *Toponymic Processes and Practices in Global View, Horizons in Geography, 2013, guest ed.; Garden Cities and Colonial Planning in Africa and Palestine,* Manchester University Press, 2014, co-ed.; *French Colonial Dakar: the Morphogenesis of an African Regional Capital*, Manchester University Press, 2016).

Sofie Boonen is a PhD Candidate at Ghent University, where she graduated as an engineer-architect. She is currently finishing her research on the spatial history of Lubumbashi, in which she traces the way the formation of the city's urban form and built environment was related to the formation of a colonial society. Together with Johan Lagae, she has published several articles on the topic.

Walter T. Brown (1943–2006) received his BS from the City College of New York and his MA and PhD from Boston University. As a Fulbright Fellow from 1967 to 1969, he conducted archival research in several European countries and field research in Tanzania for his doctoral dissertation on the history of Bagamoyo, a major nineteenth century East African trade town. During his professional life as an educator and scholar, Dr. Brown served as an assistant professor of history and director of the African Studies Center at DePauw University (Indiana) from 1970 to 1973, and then as a professor of African History and International Studies at Ramapo College of New Jersey from 1973 to 2006. An ardent advocate of study abroad for both faculty and students, he led or joined group study programmes to more than ten countries and received individual Fulbright fellowships to conduct historical research in Kenya, Colombia, and South Africa.

César Cumbe (PhD in linguistics, Université René Descartes, Paris 5, Sorbonne) is Professor and researcher at the Faculty of Language, Communication and Arts of Universidade Pedagógica, Maputo, Mozambique. He is associated with the research team on written anthropology (EHESS-Paris) and a Coordinator of Anthropological Studies at the Centre for Mozambican and Ethno-Science Studies (CEMEC-Maputo). Additional interests include linguistic and urban practices and different forms of popular expression, particularly informal inscriptions and popular art.

Hélène d'Almeida-Topor (BA, Porto-Novo, Lomé; MA, Paris 7; PhD, Paris 4) specializes in contemporary history of sub-Saharan Africa. Conducted her PhD on the economic history of Dahomey/Benin in the early colonial period, she has taught is several French universities such as the Université Paris 7 Val de Marne, Université de Metz, and Université Paris 1 Panthéon-Sorbonne. Currently a professor emeritus,

Hélène is a member of the IMAF laboratory (Paris 1/ CNRS) and also the honorary president of the *Association française d'histoire d'outre-mer*. She has published numerous articles and recently contributed to several historical dictionaries, she has also coedited several collective works on the socioeconomic history of modern sub-Saharan Africa. As an author, amongst her many books are: *Les Amazones, une armée de femmes dans l'Afrique précoloniale,* Rochevignes, 1983; *L'Europe et l'Afrique: un siècle de relations économiques*, A. Colin, 1994; and *L'Afrique au XXe siècle à nos jours*, A. Colin, 2010.

Donatien Dia Mwembu Dibwe is full professor of history at the University of Lubumbashi, DR Congo. He holds a PhD from the University of Laval, Québec (Canada). He is specialized in the social history of the mining city of Lubumbashi and authored several books on the topic, e.g. Les identités urbaines en Afrique: Le cas de Lubumbashi (Paris: l'Hamattan, 2005), Bana Shaba abandonnées par leur père: Structures de l'autorité et histoire sociale de la famille ouvrière au Katanga 1910–1997 (L'Hamattan, 2001), and, in collaboration with Rosario Giordano and Bogumil Jewsiewicki, Lubumbashi 1910–2010: Mémoire d'une ville industrielle (Paris : L'Harmattan, 2010). For several years, he presided the scientific committee of the 'Mémoires de Lubumbashi'-research group at the University of Lubumbashi.

Samira Hassa is an associate professor of languages at Manhattan College in the city of New York, United States. She holds a PhD in French Linguistics from the University of Illinois at Urbana-Champaign and a Maîtrise and a diplôme d'études approfondies (doctoral preparatory degree) from the University of Montpellier, France. She is the author of *De la Médina à la Ville Nouvelle: Etude éthnolinguistique de la ville de Fes* (2008) as well as articles on linguistic landscape and language planning. Her research interests are urban linguistics, codeswiching, and language in advertising.

Robert Home (MA, Cantab; PhD, London; DipTP, Oxford Brookes) is Professor in Land Management at Anglia Ruskin University, UK. His research interests are planning history and land management, with a strong international and African focus. His PhD was on the influence of colonial government upon Nigerian urbanisation. He has published books on British colonial town planning, land titling in Africa and the Caribbean, and planning regulations. He has also managed higher education link programmes with South Africa and Zambia, undertaken consultancies in several African countries, and supervised PhDs on African land topics.

Yossi Katz is a full professor in Geography at Bar-Ilan University, Israel. He holds the Chair for the Study of the History and Activities of the Jewish National Fund at his institution. He specializes in the modern history of the Land of Israel and the State of Israel, Zionism, the process of Jewish settlement in the Land of Israel and in Western Canada. He has published 24 books including: *Partner to Partition:*

The Jewish Agency's Partition Plan in the Mandate Era (1998); *Between Jerusalem and Hebron: Jewish Settlement in the Pre-state Period* (1999); *The Last Best West: Essays on the Historical Geography of the Canadian Prairies* (with J.L. Lehr, 1999); *The Religious Kibbutz Movement in the Land of Israel* (1999, 2003); *Inside the Ark: The Hutterites in Canada and the United States* (2012, with J.L. Lehr).

Johan Lagae is Senior Lecturer of nineteenth and twentieth century Architectural History with a particular focus on non-European regions at Ghent University. He holds a PhD on twentieth century colonial architecture in the former Belgian Congo and is a specialist of colonial and postcolonial architectural and urban history in Africa. He published widely on the topic, among others in the Journal of Architecture, Third Text, and Cahiers Africains, as well as in several edited volumes. In collaboration with Bernard Toulier and Marc Gemoets, he coauthored two monographs on the urban landscape of Kinshasa and was (co-)curator of several Congo-related exhibitions, among which The Memory of Congo: The Colonial Era (Tervuren, 2005) and, with Magnum-photographer Carl De Keyzer, Congo belge en images (Antwerp, 2010). Between 2010 and 2014, he cochaired a European funded research community on the topic "European Architecture beyond Europe".

Barbara Meiring (MA, PhD University of South Africa (UNISA)) is a Sociolinguist who taught at the Department of Afrikaans and is presently a Research Associate at the Department of Linguistics at UNISA. Was involved at the United Nations Training Course in Toponymy for the Africa South Division, she is a member of the National South African Geographical Names Council, and was a member of the Southern African Association for Research and Development in Higher Education until retirement in 1996. As a member of the Names Society for Southern Africa, she served as Editorial Secretary for *Nomina Africana,* the Journal of the Society of which she is now an Honorary Member. As Onomastician, she publishes articles on various aspects of geographical names annually and continues to deliver presentations at various international symposia and conferences. She (co-)authored books like *Funksionele Afrikaans* (Van Schaik, Pretoria, 1991) and *Leer Presteer* (Southern Publishers, Pretoria, 1991) and is the editor of *Guidelines for the Standardisation of South African Geographical Names* to incorporate all eleven National languages. This document is presently situated at the National Department of Arts and Culture as a working document being updated regularly in consultation with the different Language Boards.

Garth Myers is the Paul E. Raether Distinguished Professor of Urban International Studies at Trinity College, Hartford, Connecticut. Myers has taught at the University of Kansas, University of Nebraska-Omaha, Miami University (Ohio), California State University at Dominguez Hills, and UCLA. He is the author of three books (*African Cities, Disposable Cities,* and *Verandahs of Power*) and more than 60 articles and book chapters, as well as coeditor of two other volumes. Almost all of his publications focus on the political, cultural, and historical geography of urban development in Sub-Saharan Africa.

Ambe J. Njoh is Professor/Director of Urban and Regional Planning at the School of Urban Affairs, and School of Geosciences, University of South Florida, USA. He has written eleven books and published more than 60 articles in peer-reviewed journals. His most recent book, *Urban Planning and Public Health in Africa*, was published by Ashgate (2013). He just completed the manuscript of a book on *French Colonial Urbanism* slated to be published by Routledge.

Gordon H. Pirie (MA, PhD, University of the Witwatersrand) is Professor of Geography and Environmental Studies at the University of the Western Cape, South Africa, and the Deputy Director of the African Centre for Cities, University of Cape Town. He has taught and published extensively on African transport and travel on various scales and times: geographical, social, ecological, and epistemic issues. On the editorial boards of the *Journal of Transport History* and the *Journal of Transport Geography*, Pirie is a founding member of the International Geographical Union's Commission on Transport and Geography. Amongst his recent books are: *Cultures and Caricatures of British Imperial Aviation: Passengers, Pilots, Publicity* (Manchester University Press, 2012); *Air Empire: British Imperial Civil Aviation, 1919–1939* (Manchester University Press, 2009); and *Mobility in History: the State of the Art in the History of Transport, Traffic and Mobility* (Éditions Alphil, 2009).

Chapter 1
Introduction: Place Names in Africa: Colonial Urban Legacies, Entangled Histories

Liora Bigon

Abstract Enframing the collective volume, this chapter elaborates on the distinctiveness of this volume and its contribution to recent critical place-names studies in terms of aims, scope, geography and methodology. It also provides an in-dept historiographic analysis in the field of toponymy against the background of (post-)colonial urban environments in Africa. Implying on a variety of toponymic inter-crossings between time spans, spatialities, scales and (post-)colonial legacies, the advantage of using the relational analytic approach of the 'entangled' in this context is brought into light.

Keywords African toponymy • Urbanism • (Post-)colonial history • Colonial heritage • Place-names historiography • Colonial space • Pre-colonial settlement

This volume examines the discursive relations between indigenous, colonial and post-colonial legacies of place-naming in Africa in terms of the production of urban space and place. It is conducted by tracing and analysing place-naming processes, particularly in sub-Saharan Africa during colonial times (British, French, Belgian, Portuguese), with a considerable attention to both the pre-colonial and post-colonial situations. Quite untypically of both traditional and recent place naming studies, which have been governed by linguists and historians of (European) nationalism respectively, this volume is primarily a product of urbanists – that is, of urban historians, planners, geographers and architects. With a strong bias towards the physical environment and the many ways in which it is perceived, this collective work expands on several aspects such as: site and street names; generic and specific names; top-down and bottom-up naming systems, ranging from the highly political to the mundane; toponymic tapestries and simultaneous naming procedures; and names in the arena of intra-national and international relations. Our analysis also focuses on several key issues that faithfully converse with the multiplicity of colonial legacies in the continent, amongst them: the influence of colonial doctrines

L. Bigon (✉)
The Institute of Western Cultures, The Hebrew University of Jerusalem, Jerusalem, Israel

Holon Institute of Technology, Holon, Israel
e-mail: liorin@hotmail.com

L. Bigon (eds.), *Place Names in Africa*, DOI 10.1007/978-3-319-32485-2_1

on colonial regional and urban planning and the consequent spatial terminologies; the interaction of autochthonous nomenclatures, some of which are very old yet nonetheless strong and persistent, with the modern strata of colonial lexical logic, as well as post-colonial interactions; and the ideological dimension, or its total absence, in formal and informal place naming as a reflection of regional, national or ethnic identities, languages, memories and counter-memories. An investigation of these issues advances our understanding of the very ways by which naming corpuses are inscribed into the urban text (signposted-cum-oral), and sub-text (visual-cum-perceptual), and range between complementary, competitive and highly contested.

By combining in-depth area studies research – some of the contributions are of ethnographic quality – with colonial history, planning history and geography, we intend to show that culture matters in research on place names. We shall go beyond the recent understanding obtained in critical studies of nomenclature, normally based on lists of official names, that place naming reflects the power of political regimes, nationalism, and ideology. While these aspects are carefully scrutinised in this volume, a broader and more inclusive long-term approach is generally embraced, giving precedence to historical naming processes and accompanying cultural structures. These processes and structures are conceived here as a rich source for analysis of the more nuanced nature of power and wider socio-cultural conditions, using a variety of primary and secondary sources, including multi-site archival research and field experience. The present collection is also a pioneer in aiming to fill in some historiographic gaps related to the place of the global South in general and (sub-Saharan) Africa in particular in the burgeoning field of place-name studies, which are mostly Eurocentric; and the place of place-name studies in literatures of (post-) colonialism, cultural geography, urban and architectural planning, and African histories.

1.1 Why Africa? Place-Name Historiographies, Methodologies

Research tendencies in place-name studies (toponymy) in anglophone, francophone and lusophone Africa are a mirror of the wider trends in toponymic research. The post-1990s period is characterised as a 'critical turn' in place-name scholarship, excitingly producing in its "explicit and self-reflexive engagement with critical theories of space and place" a "new body of research, which situates the study of toponymy within the context of broader debates in critical human geography."[1] Yielding, however, only a meagre number of references concerning

[1] Rose-Redwood, Reuben, Derek Alderman and Maoz Azaryahu, 'Geographies of Toponymic Inscription: New Directions in Critical Place-Name Studies', *Progress in Human Geography*, 34, 4 (2010), pp. 453–470 (p. 455).

sub-Saharan Africa,[2] this period is clearly different from the more antiquarian and taxonomical approaches featured in most pre-1990s research. The latter was governed by linguists and topographers rather than by historians, human geographers or political scientists. Thus, its concerns remained somewhat parochial, limited to toponymic etymology and typology, with hardly any interdisciplinary influence. Such approaches characterised contemporary research on toponyms (and ethnonyms) in sub-Saharan Africa as well.[3]

It has been clearly acknowledged in a recent critical study of place names that beyond the primary purpose of place names as an administrative act designated to facilitate spatial orientation – the symbolic and socio-political aspects of their production must be also be taken into account. Yet on the other hand, in the current literature on urban management and governmentality in the global South, featured in a 'developmentalist' viewpoint concerning Africa in particular, scant attention is accorded to the meaning of place names and their symbolism. More qualitative and technical issues as to efficient spatial navigation in African cities are clearly preferred to quantitative ones.[4] This partly explains the existence of several concurrent toponymic systems in certain African cities; and the irrelevancy for most

[2]These references are therefore worthy to be mentioned in detail: Hélène d'Almeida-Topor, 'Le modèle toponymique colonial dans les capitales de l'ouest africain francophone', in Catherine Coquery-Vidrovitch and Odile Goerg (eds), *La ville européenne outre mers* (Paris: l'Harmattan, 1996), pp. 235–243; Liora Bigon, 'Names, Norms and Forms: French and Indigenous Toponyms in Early Colonial Dakar, Senegal', *Planning Perspectives*, 23 (2008), pp. 479–501; Liora Bigon and Ambe Njoh, 'The Toponymic-Inscription Problematic in Urban Sub-Saharan Africa: From Colonial to Postcolonial Times', *Journal of Asian and African Studies*, 50, 1 (2015), pp. 25–40; James Duminy, 'Street Renaming, Symbolic Capital, and Resistance in Durban, South Africa', *Environment and Planning D: Society and Space*, 32 (2014), pp. 310–328; Garth A. Myers, 'Naming and Placing the Other: Power and the Urban Landscape in Zanzibar', *Tijdschrift voor Economische en Sociale Geografie*, 87, 3 (1996), pp. 237–246; Ambe J. Njoh, 'Toponymic Inscription, Physical Addressing and the Challenge of Urban Management in an Era of Globalization in Cameroon', *Habitat International*, 34 (2010), pp. 427–435; Tiago Castela and Maria Paula Meneses, 'Naming the Urban in Twentieth-Century Mozambique: Towards Spatial Histories of Aspiration and Violence', lecture presented at the international conference 'Colonial and Postcolonial Urban Planning in Africa', the IPHS and the Institute of Geography and Spatial Planning, University of Lisbon, 5–6 September 2013.

[3]For only few examples see: Ikany Burije, 'Note sur l'orthographe des principaux noms géographiques du Burundi', *Kongo-Overzee*, 23, 3–4 (1957), pp. 224–225; Enrico Cerulli, *Per la toponomastica della Somalia* (Roma: Instituto per l'Oriente, 1931); Louis-Ferdinand Flutre, *Pour une étude de la toponymie de l'AOF* (Dakar: Université de Dakar, Faculté des lettres, 1957); Claude Gouffe, 'Problèmes de toponymie Haoussa: les noms de villages de la région de Maradi (République du Niger)', *Revue internationale d'onomastique*, 2 (1967), pp. 95–127; Eugene Kirchnerr, *Place Names of Africa, 1935–1986: A Political Gazetteer* (Netuchen: Scarecrow, 1987); Archibald Tucker, 'Conflicting Principles in the Spelling of African Place Names', *Onoma* 7, 2 (1956/7), pp. 215–228.

[4]José Anson, *Connecting the 'Unconnected' in Sub-Saharan Africa: Postal Networks can Leverage Access to Infrastructure* (Bern: United Nations' Universal Postal Union, 2007); Serena Coetzee and Antony Cooper, 'The Value of Addresses to the Economy, Society and Governance: A South African Perspective', Paper for the 45th Annual Conference of the Urban and Regional Information System Association (URISA), Washington DC, 20–23 August 2007; Catherine

of the urban residents of top-down systems introduced by exogenous programmes for economic development, such as the World Bank's (numerical) street addressing programme.[5]

While the recent critical study of place names does expand on the symbolic and socio-political aspects of the production of the names, it is over-concerned with the understanding that place naming reflects the power of modern political regimes, nationalism, and ideology. The preoccupation with political power's control over both landscape and history is especially true for publications in English, which tend to be centred on the West (North America and Western Europe) and Eastern Europe,[6] with only a few geographic exceptions.[7] Moreover, some of the relevant encyclopaedic entries and other key references are highly Eurocentric[8]; and critical, non-monograph book-length studies are also rare.[9] As summarised by Kenneth

---

Farvacque-Vitkovic, Lucien Godin, Hugues Leroux and Roberto Chavez, *Street Addressing and the Management of Cities* (Washington DC: The World Bank, 2005).

[5]For the World Bank's programme for sub-Saharan Africa see Farvacque-Vitkovic (et al), *Street Addressing*. For its critique and the existence of simultaneous naming systems see: Njoh, 'Toponymic Inscription, Physical Addressing'; Bigon and Njoh, 'The Toponymic-Inscription Problematic'; Hélène d'Almeida-Topor's contribution in this volume.

[6]For a representative though incomplete list (in alphabetical order) see: Maoz Azaryahu, 'German Reunification and the Politics of Street Names: the Case of East Berlin', *Political Geography*, 16, 6 (1997), pp. 479–493; Rhys Jones and Peter Merriman, 'Hot, Banal and Everyday Nationalism: Bilingual Road Signs in Wales', *Political Geography*, 28 (2009), pp. 164–173; Duncan Light, Ion Nicolae and Bogdan Suditu, 'Toponymy and the Communist City: Street Names in Bucharest, 1948–1965,' *GeoJournal*, 56 (2002), pp. 135–144; Daniel Milo, 'Street Names,' in Pierre Nora (ed.), *Realms of Memory* (New York: Columbia University Press, 1997), pp. 363–389; John Murray, *Politics and Place-Names: Changing Names in the Late Soviet Period* (Birmingham: University of Birmingham, 2000); Catherine Nash, 'Irish placenames: Post-Colonial Locations', *Transactions of the Institute of British Geographers*, 24, 4 (1999), pp. 457–480; Paulina Raento and Cameron Watson, 'Gernika, Guernica, *Guernica*? Contested Meanings of a Basque Place', *Political Geography*, 19 (2000), pp. 707–736; Reuben Rose-Redwood, 'From Number to Name: Symbolic Capital, Places of Memory, and the Politics of Street Renaming in New York City', *Social & Cultural Geography*, 9, 4 (2008), pp. 432–452.

[7]Among the most prominent here are the cases of (post-)colonial Singapore and Israel/Palestine, where critical toponymic study has been carried out by native scholars exclusively. Brenda Yeoh, 'Street Names in Colonial Singapore', *Geographical Review*, 82, 3 (1992), pp. 313–322; Brenda Yeoh, 'Street Naming and Nation-Building: Toponymic Inscriptions of Nationhood in Singapore', *Area*, 28, 3 (1996), pp. 298–307. For references on Israel/Palestine see notes 20–24.

[8]For instance, the entry 'Street Names and Iconography' (vol. 10, pp. 460–465) in the *International Encyclopedia of Human Geography*, ed. by Rob Kitchin and Nigel Thrift (Oxford: Elsevier, 2009) – comprises no toponyms outside of the global North! Similarly, the entry 'Semiotics' in this same source (vol. 10, pp. 89–95), heavily designed around toponymic issues, mentions only three toponyms from the global South, out of a total of sixteen mentioned. For similar rates see: Maoz Azaryahu, 'The Power of Commemorative Street Names', *Environment and Planning D: Society and Space*, 14 (1996), pp. 311–330; Naftali Kadmon, *Toponymy : The Lore, Laws and Language of Geographical Names* (New York: Vantage Press, 2000).

[9]The edited volume by Lawrence Berg and Jani Vuolteenaho entitled *Critical Toponymies: The Contested Politics of Place Naming* (Surrey: Ashgate, 2009) – is an exception by being a critical collection in English with global geographic coverage. Together with this, the collection reflects

Foote and Maoz Azaryahu in their article surveying contemporary research in geography on issues of public memory and commemoration as expressed materially in landscape, memorials and heritage sites: "[m]uch of the research both inside and outside of geography has focused on wars, revolutions, and other major historical events [in Europe and North America] from the late eighteenth, nineteenth and twentieth century." Amongst the issues that have received the most attention "|are nationalism and wars of independence." This means, they argue, "that scholarship is rich for some periods and places, but weak in others particularly for Latin American, Asia, and Africa."[10]

A unique project that challenges this situation in its time span, theoretical, thematic and geographic scope is a monumental collection published by UNESCO under the direction of the French sociologist Christian Topalov.[11] Entitled *Les divisions de la ville* and conceptualised as part of the collection 'Les mots de la ville' (words of the city), it is aimed at the examination of the social processes that structure urban agglomerations through their spatial terminologies. Each contribution in this series of case studies typically deals with an individual city in order to explore the local dimension in the creation of lexical corpuses, while a thorough comparative analysis has been prepared by Topalov. Apart from a welcome emphasis on the southern hemisphere in the selected cities for discussion,[12] the tripartite thematic organisation of the collection (from the *Anciens Régimes* to contemporary modernisation via the shock of western expansion) is outstanding in giving precedence to the historical factor. Embracing a *longue durée* approach, it is presumed that without historical depth the understanding of the specific connection between language, territory and society is impossible. Therefore, Topalov's main interest is only complementary to the overall implemental target of UNESCO in this series, which is to encourage innovative research for the amelioration of urban management and the creation of public politics. According to him: "in capturing the words for an object, one can better understand the [actual, perceptual] divisions

---

the thematic mainstream of the post-1990s research aiming to tackle place naming as "political practice par excellence of power over space" (p. 1); analysis and conclusion are left to the readers; and it seems that all chapters but the editors' Introduction had been already published. See also: Reuben Rose-Redwood and Derek H. Alderman, Guest Editors, Special Thematic Interventions Section: 'New Directions in Political Toponymy', in *ACME: An International E-Journal for Critical Geographies*, 10, 1 (2011), http://www.acme-journal.org/volume10-1.html (visited 1 July 2015).

[10]Kenneth Foote and Maoz Azaryahu, 'Toward a Geography of Memory: Geographical Dimensions of Public Memory and Commemoration', *Journal of Political and Military Sociology*, 35, 1 (2007), pp. 125–144 (pp. 125, 126).

[11]Christian Topalov (ed.), *Les divisions de la ville*, collection 'Les mots de la ville', prepared under UNESCO's Management of Social Transformations (MOST) Programme (Paris: UNESCO, Maison des sciences de l'homme 2002).

[12]Governed by socio-linguists and historians, the twelve contributions in the collection consist of two cases each from North Africa, West Africa, South America and the Far East, and one on India. Europe is represented by two English cities and Italy. Interestingly, Eastern Europe, North America and Australia are not represented, nor are anglophone, East or southern Africa.

of cities. Beneath the prominent simplicity of spatial divisions of the modern administration, the traces of ancient institutions can be perceived, the placement of the past within the present, and the spatial claims of groups."[13]

The present collection is more in line with the francophone tradition in critical place-names studies,[14] which invites us to a voyage in space and time, in languages and disciplines. It contributes to the growing recognition that the naming of streets and urban sites is also involved in questions that go beyond the |nationalistic. This recognition has been crystallised vis-à-vis both the preoccupation of anglophone toponymic historiography with nationalism and the actual situation in post-colonial sub-Saharan Africa in this regard. The latter situation indicates that the ideology of African nationalism and the strategy of nation building "|rest on a vision of the future, rather than on past memories or current reality"; and, with marginal historical relevance, nation building is "not only a hopeless effort, but also a most damaging one."[15] We shall therefore embrace a broader and more inclusive long-term perspective, with considerable attention given to historical toponymic processes and related cultural structures. In other words, by examining what is problematic and inherent in some past and present toponymic references in Africa, "[r]ather than simply reflecting the impress of state or elite ideologies", to borrow the words of Garth Myers on Zanzibar, we seek to show how such toponyms "indicate the presence of many interwoven layers of power."[16] Thereby, these toponyms – many of which are located between exogenous and indigenous models, regional and urban, formal and informal, generic and specific – constitute a rich source for analysis of the 'capillary' nature of power[17] and wider socio-cultural processes.

Against this background, we can ask: is there African urban toponymy or toponyms in urban Africa? While intellectual debates concerning African distinctiveness continue, we do not believe that there is anything exceptional about the continent. Yet understanding the local, regional, national, international, transnational, pre-colonial, colonial and post-colonial contexts of place names in Africa and its urban environment might be useful. This is because it encourages some

[13]Topalov (ed.), *Les divisions de la ville*, p. 1 (Editor' translation).

[14]See also: Hervé Guillorel (ed.), *Toponymie et politique: les marqueurs linguistiques du territoire*, la collection 'Droits, territoires, cultures' (Bruxelles: Bruylant, 2008); Hervé Guillorel (ed.), a special issue on 'Onomastique, droit et politique', *Droit et Cultures: Revue internationale interdisciplinaire*, 64, 2 (2012); and the session 'What's in a Name? How We Label Peripheral Places' at the European Association for Urban History 11th International Conference on Urban History, Prague, 31 August 2012 (including Topalov's lecture on 'The Urban Vocabulary of Social Stigma in Late 20th Century France').

[15]Mordechai Tamarkin, 'Nationalism and Nation-Building in Black Africa: Fateful Connections', *Canadian Review of Studies of Nationalism*, 20, 1–2 (1994), pp. 83–91 (p. 84). See also: Mordechai Tamarkin, Culture and Politics in Africa: Legitimizing Ethnicity, Rehabilitating the Post-Colonial State', *Nationalism and Ethnic Politics*, 2, 3 (1996), pp. 360–380.

[16]Myers, 'Naming and Placing the Other', p. 237.

[17]Myers, 'Naming and Placing the Other', p. 237.

new insights into comparative or inter-crossed themes from within (and beyond) the continent, also methodologically (see below). Toponymic situations in practice and theory which challenge some typical western conceptions are inspiring, and many of these are discussed by the contributors to this volume, for example: fluid borderlands; mobile spaces and changing place names; multiple towns in a region bearing the very same name; multiple streets in the same vicinity bearing the same name; toponym that is meaningful in multiple languages; local use of an alternative toponym that is off-the-radar of an officially recognised name on the national and international levels; toponymy reflecting an indigenous ecosystem; the ways by which European toponymy outside Europe was twisted to meet colonial conditions and mentalities; the importance of both never-used suggested names and names void of etymological meaning; the re-introduction of colonial names in the post-independence era in spite of toponymic Africanisation; post-colonial toponymic transformation in modern multi-ethnic societies; and the disturbance of international relations by means of toponymic conflict.

The methodological question is conspicuous in this context. We strongly believe that studying place names in sub-Saharan Africa can significantly enrich place-name studies beyond this region in general and in the global North-West in particular. A classical methodological problem in post-1990s research is the inadequacy of "the reliance on maps and gazetteers to study place names" without being "supplemented with some combination of archival research, participant observation, interviews, and ethnographic methods."[18] Some research is manifestly unidirectional, such as that on street-renaming policies in European cities following revolutionary changes of political regimes. Relying on maps and street plans, it is impossible to assess to which extent these top-down modes of renaming were absorbed in situ by the urban residents, if at all, or if they evoked any response at all.[19]

The arrogance of some other research is simply astounding. In the case of the highly ideological and contested environment of Israel/Palestine, for instance, the Hebraisation of the map of the land generates constant conflict with indigenous toponyms it proposes to erase, change, or replace. As noted by Meron Benvenisti, the policy of the Israeli Names Committee always involved a triangular oscillation between the creation of an 'historical map', that is, using only Arab names as an epistemological channel to identify ancient Hebrew names and thus 'redeem' the maximum number of ancient names; an 'actual map', namely an updated map without the names of Arab settlements and other landscape objects relevant to the pre-1948 period; and a 'purely Hebrew map' that includes, inter alia, a transcription that reflects Hebrew toponymic pronunciation.[20] Indeed, it seems that the ongoing reality of the Jewish-Arab conflict in Israel/Palestine has engendered not only a

[18]Rose-Redwood (et al.), 'Geographies of Toponymic Inscription', p. 466.

[19]Light (et al.), 'Toponymy and the Communist City'; Azaryahu, 'German Reunification.'

[20]Meron Benvenisti, 'The Hebrew Map', *Theory and Criticism*, 11 (1997), pp. 7–29 [in Hebrew]. For an English version see: Meron Benvenisti, *Sacred Landscape: The Buried History of the Holy Land since 1948*, trans. M. Kaufman-Lacusta (Berkeley: University of California Press, 2000).

divided and split space along status, ethnic and national lines, but also split place-name historiographies.

The bulk of critical research on toponymy produced by Israeli geographers has been preoccupied almost solely with developments within the Jewish sector.[21] When Palestinian toponymy has been discussed under this conceptual mechanism, there has again been an over-reliance on towns maps without any field observation or knowledge of Arabic.[22] One article even stated that the "most prominent group of Arab settlements are those where it is impossible to assign the origin of their names because the various gazetteers did not provide the origin of the name or rather disagreed on its origin."[23] This calls for an acquaintance with the Palestinian semiotic reservoir of thousands of place names and geographical objects, and hundreds of explanations about their origins and the reasoning behind them.[24] This somewhat colonialist laissez-faire approach also plays with toponymy on the ground: Israeli road signs in many cases are ambiguous for Palestinians,[25] and municipal signposting is sometimes careless towards the subaltern: a signpost for 'Brazzaville Street' in Holon, a Tel Aviv suburb, states that this city is the "capital of Congo that is in EAST Africa" (Editor's emphasis) (Fig. 1.1). This ignorance might be both explained in Israel's innocence of administering a global imperial past, and in its current administration of a colonial present.[26]

---

[21]For a partial yet representative list of publications in English see: Maoz Azaryahu and Arnon Golan, '(Re)naming the Landscape: The Formation of the Hebrew Map of Israel, 1949-1960', *Journal of Historical Geography*, 27, 2 (2001), pp. 178–195; Maoz Azaryahu and Aharon Kellerman, 'Symbolic Places of National History and Revival: A Study in Zionist Mythical Geography', *Transactions of the Institute of British Geographers*, 24 (1999), pp. 109–123; Yoram Bar-Gal, 'Naming City Streets: A Chapter in the History of Tel Aviv, 1909-1947', *Contemporary Jewry,* 10, 2 (1989), pp. 39–50; Yossi Katz, 'Identity, Nationalism and Place Names: Zionist Efforts to Preserve the Original Local Hebrew Names in Official Publications of the Mandate Government of Palestine', *Names* 43, 2 (1995), pp. 103–118; Yossi Katz, 'Reclaiming the Land: Factors in Naming the Jewish Settlement in Palestine during the Era of the British Mandate', in *These are the Names: Studies in Jewish Onomastics*, ed. by Aaron Demsky (Ramat Gan: Bar-Ilan University Press, 1999), vol. II, pp. 63–112.

[22]Maoz Azaryahu and Rebecca Kook, 'Mapping the Nation: Street Names and Arab-Palestinian Identity: Three Case Studies', *Nations and Nationalism*, 8, 2 (2002), 195–213; Nurit Kliot, 'The Meaning of the Arab Settlements' Names in Eretz Israel and their Comparison to Hebrew Settlements' Names', *Horizons in Geography*, 30 (1989), pp. 71–79 [in Hebrew]. Methodologically exceptional is: Noam Shoval, 'Street-Naming, Tourism Development and Cultural Conflict: the Case of the Old City of Acre/Akko/Akka', *Transactions of the Institute of British Geographers*, 38 (2013), pp. 612–626. In Palestinian popular maps Israeli settlements are non-existent.

[23]Kliot, 'The Meaning of the Arab Settlements' Names', p. 72.

[24]Amer Dahamshy, 'A Name for a Place: The Naming of Arab Settlements and Natural Features in the Galilee's Folk Narratives' (PhD thesis, the Hebrew University of Jerusalem, 2009) [in Hebrew].

[25]Liora Bigon and Amer Dahamshy, 'An Anatomy of Symbolic Power: The Israeli Road-Signs Policy and the Palestinian Minority', *Environment and Planning D: Society and Space*, 32, 4 (2014), pp. 606–621.

[26]Derek Gregory, *The Colonial Present: Afghanistan, Palestine, Iraq* (Oxford: Blackwell, 2004).

Fig. 1.1 'Brazzaville Street', Holon, Israel (Editor's photo)

Stressing the methodological point, it is not therefore surprising that it was an Africanist geographer who criticised Cohen and Kliot's work on Israeli place names as a powerful tool for reinforcing competing nationalistic ideologies[27]: "their analysis is an account of how names have been used to reinforce Zionism; little is made of Palestinian resistance. It is crucial to look to all sides in order to truly see the 'fields of power' at work in the production and reproduction of space."[28] While most of the recent place name scholarship is no longer blind to the possibility of resistance and even "risks becoming a bit too predictable and formulaic in its repetitions invocations of toponymic domination and resistance"[29] – Africanist geographers, urbanists and historians have been essentially pioneers in opening a rich methodological spectrum.

This involves working across a variety of possible disciplines and sources, with Africanists traditional reliance on archaeology, anthropology, comparative linguistics, visitors' accounts and oral traditions. This comes in the absence of historical indigenous written sources in most of sub-Saharan Africa in pre-colonial times (Islamic literacy and then European literacy are limited to certain areas and

[27]Saul B. Cohen and Nurit Kliot, 'Place Names in Israel's Ideological Struggle over the Administered Territories', *Annals of the Association of American Geographers*, 82, 4 (1992), pp. 650–680.

[28]Myers, 'Naming and Placing the Other', p. 238.

[29]Rose-Redwood (et al.), 'Geographies of Toponymic Inscription', p. 466.

eras).[30] Because of the inter-dependence of these sources, working with different source types and disciplines contributes to the reliability of the results. As concluded by Stephan Bühnen in his comparative study on the usability of toponymy in Africa for historical research: "I have tried to reconstruct the original place names, made attempts at their interpretation, proposed a relative date for their genesis, and showed that their spatial distribution permits historical interference (e.g., about culture change processes). The place names took us back into a time not covered by written sources."[31]

Anthropological research is also usable here, proving inter alia that some settlement conceptions of residents are not necessarily about a fixed physical location on a map, but rather about membership of a community that might include spiritual ancestors. The occupants of a Zimbabwean settlement, for instance, did not feel 'at home' in its physical area, as they believed that their officially appointed chief had been selected without consulting them. In striving for a chosen chief with common ancestral links, they planned to re-establish their settlement elsewhere. This evokes the concept of 'shifting settlements', where attempts at cartographically-fixed nomenclature ('one place, one name') might even constitute a pitfall.[32] Similarly, the conjunction of a variety of early written and cartographic sources in Arabic and a variety of European languages has pointed to the usability of toponymy as a source of information for revealing the African past. Many hundreds of maps and charts – produced over the course of centuries and multilaterally spread over numerous archives – transcend the exclusive preserve of cartography scholars.[33]

Particularly deserving of attention here are the papers of the expert meeting held in Paris by UNESCO in 1978 entitled *African Ethnonyms and Toponyms*[34] – a collection launched out of an urgent methodological need to establish a standardised transcription of naming systems for the continent. In fact, this collection had been conceptualised as a direct response to the onomastic difficulties raised in the preparation of another UNESCO collection, the monumental eight volume

[30]Written sources for sub-Saharan Africa history (apart from East Africa) gradually emerged with the introduction of Arabic script and accounts from the ninth century – however, these are confined to the Sahel and the northern savannah belt. European accounts originated in the mid-fifteenth century, first for the Senegambia region, but were limited to the coast and its environs, with some ambiguous and often indirect references to the hinterland.

[31]Stephan Bühnen, 'Place Names as an Historical Source: An Introduction with Examples from Southern Senegambia and Germany', *History in Africa*, 19 (1992), pp. 45–101 (p. 90).

[32]Rowan Roenisch, 'Mental Maps and Shifting Settlements: The Invisible Boundaries of the Zimbabwean Shona and Tonga Home', *Traditional Dwelling and Settlements*, 144 (2002), pp. 17–54 (pp. 25–26).

[33]René Baesjou, 'The Historical Evidence in Old Maps and Charts of Africa with Special Reference to West Africa', *History in Africa*, 15 (1988), pp. 1–83; Ann McDougall, 'The Quest for "Tarra": Toponymy and Geography in Exploring History', *History in Africa*, 18 (1991), pp. 271–289.

[34]UNESCO, *African Ethnonyms and Toponyms* (Paris: UNESCO, 1984).

project of the *General History of Africa* (1964–1999). It is unknown whether the European experts of *African Ethnonyms and Toponyms*, who embraced a softer critical approach towards colonial transcriptions, and the African experts, who were more militant in their call for the 'decolonisation' of names, ever reached a consensus on a system of transcription and whether such a system was capable of application.[35]

However, some of the papers adopted an explicitly functionalistic methodological standpoint, such as that of the Senegalese scholar Cheikh Anta Diop, who investigated population movements as explanations for the origin of man. In the absence of archaeological data or written sources, Diop used linguistic, ethnic (ethnonyms and toponyms) and socio-political data to reach virtual certainty in the study of Africa's past. By a comparative analysis of linguistic relationships between two geographically remote groups of languages (Ancient Egyptian, Coptic and the Senegalese Wolof), the existence of cultural contacts between the Nile Valley and the West African coast were demonstrated.[36] Indeed, due to contemporary scientific prejudice, Diop, an historian, anthropologist, physicist and politician who submitted several PhD dissertations in Paris, could not find a jury of examiners for his thesis on the subject referred to above, completed in 1954.

Thematically, some of the works on Africa's toponyms and especially the pre-1990s research has never broken out from the conceptual and geographic limits of 'Area Studies.' Yet such eminent methodological creativity on the part of Africanist scholars should have informed place names scholarship beyond Africa and widened its horizon. In many respects, Bühnen's designation: "la toponymie [ouest-]africaine n'est pas encore sortie de l'oeuf",[37] the process of shedding the egg shells, also in terms of methodology – refers not so much to the remote past but rather to the (post-) colonial urban environments in Africa as well. Toponymic inscriptions are not always inscribed into the city, and urban spatial terminology needs to be decoded as text, without the obligatory presence of written text (street signs, accurate city maps). Such an approach promotes seeing the landscape as text rather than looking for texts inscribed into the landscape,[38] for 'nondescript spatial structures' characterise many cities in Africa and generally in the global South.

[35]See for instance: Pierre Alexandre, 'Some Problems of African Onomastics' (pp. 51–67) and David Dalby, 'The Transcription of Ethnonyms and Toponyms in Africa in Relation to their Historical Study' (pp. 80–85) versus Olabiyi Yai, 'African Ethnonymy and Toponymy: Reflections on Decolonization' (pp. 39–50) – all in UNESCO, *African Ethnonyms and Toponyms.*

[36]Cheikh Anta Diop, 'A Methodology for the Study of Migrations', in UNESCO, *African Ethnonyms and Toponyms*, pp. 86–109. See also: Cheikh Anta Diop, *L'Afrique noire précoloniale* (Paris: Présence Africaine, 1960).

[37]Bühnen, 'Place Names as an Historical Source', p. 45.

[38]This argument is inspired by: Jim Duncan and Nancy Duncan, '(Re)reading the Landscape', *Environment and Planning D: Society and Space*, 6 (1988), pp. 117–126; Jack Goody, *The Interface Between the Written and the Oral* (Cambridge: Cambridge University Press, 1987).

Streets are nameless, have multiple names or have names that are not signposted, and buildings are not necessarily numbered.[39] Our work in the city centre of Dakar, Senegal, for example, revealed an alternative naming system being used informally by the Lébou residents in parallel to the French street-naming system for the urban grid[40]; and our fieldwork in Cameroon provides further glimpses into such an ambiguity. The town Limbé on the Atlantic Ocean was named 'Victoria' at its establishment in 1858 by a British Baptist missionary in honour of the Queen of England. It went under that name from its founding to 1982. Despite its colonial roots, some of the older generation from the region still prefer to refer to the town as 'Victoria.'[41] Similar situations of using colonial names in spite of post-independence Africanisation/Arabisation of names were also noticed in modern Dakar and even in Oran – a case of a white settler colony where one might expect a more extremist reaction.[42] This blurs the assertion that place names maintain ideological attachments to places,[43] yet does not contradict more passionate situations as implied in the studies of Samira Hassa or Wale Adebanwi in this collection.

Despite Limbé's neat orthogonal street layout, however, only a few streets are signposted (Fig. 1.2). Yet, from autochthonous viewpoint, there is hardly any nameless street throughout the city. In the same region, Mutengene, a major crossroads town with about 70,000 residents, consists of neither a neatly configured street pattern nor paved streets. The buildings bear hand-painted letters and numbers (Fig. 1.3), inscribed by the local electrical power provider for billing purposes. It has six major streets, named after salient objects (such as 'electric line') or orientation towards neighbouring-towns, as well as several minor streets, none of which is signposted.[44] Where the locals operate in an unmistakably well-defined space, a westerner may experience chaos and disorder.

This prevalence of nondescript toponymic addresses and spatial codification systems in urban sub-Saharan Africa, is often quite difficult to detect by a visitor,

---

[39] Ambe Njoh, *Planning in Contemporary Africa: The State, Town Planning and Society in Cameroon* (Aldershot: Ashgate, 2003), pp. 235–260.

[40] Liora Bigon, 'Names, Norms and Forms: French and Indigenous Toponyms in Early Colonial Dakar, Senegal', *Planning Perspectives*, 23 (2008), pp. 479–501. For a similar co-existence of the Maori system of naming and mapping with the new one imposed by colonial surveyors see: Giselle Byrnes, 'Surveying – The Maori and the Land: An Essay in Historical Representation', *New Zealand Journal of History*, 31, 1 (1997), pp. 85–98.

[41] Bigon and Njoh, 'The Toponymic-Inscription Problematic', p. 36.

[42] Respectively: Bigon, 'Names, Norms and Forms', p. 496; Belkacem Boumedini and Nebia Dadoua Hadria, 'Les noms des quartiers dans la ville d'Oran: entre changement official et nostalgie populaire', *Droit et Cultures*, 64, 2 (2012), pp. 149–160. In Aotearoa, New Zealand, also a case of white settler colony, the reapplication of selected Maori names by the settlers may be seen as an attempt in the structuring of colonialist cultural and national identities. See Giselle Byrnes, '"A Dead Sheet Covered with Meaningless Words?": Place Names and the Cultural Colonization of Tauranga', *New Zealand Journal of History*, 36, 1 (2002), pp. 18–35 (p. 29).

[43] Kadmon, *Toponymy*, p. 79.

[44] Bigon and Njoh, 'The Toponymic-Inscription Problematic', p. 36.

Fig. 1.2 A view of the orthogonal street layout of Limbé. Notice the absence of any official street signage (Editor's photo)

but has been also skilfully portrayed by the architectural historian Johan Lagae with regards to Lubumbashi's unofficial linguistic-visual landscape:

> Many private parcels indeed house bars, or 'ngandas' as they are locally called, without any exterior sign explicitly signaling the more public status of the site. On my very first visit in Lubumbashi, I remember drinking beer in the living room of a colonial villa of a complete stranger, the bar-function of the house only being subtly indicated by the presence of an empty beer bottle on top of the garden wall. It was a sign I would have missed if it hadn't been pointed out to me by my local informant of the time, who explained its underlying logic: the everyday struggle for survival in an economically harsh context requires inventive ways of procuring revenue without attracting too much attention, especially not that of the local police force which is always eager of earning a couple of hundreds of Congolese francs or getting drinks for free.[45]

Such topomymic ambiguity and many other trajectories, in which a variety of historical contexts are embedded, necessitate an intimate acquaintance with each of this volume's participants (eight out of seventeen participants are natives to Africa) and the respective studied site/s, relevant periods, languages, and primary and other sources. They also necessitate a conscious employment of innovative research

[45] Johan Lagae, 'Navigating [Off Radar": Liminal Spaces in the City Centre of Colonial/Postcolonial Lubumbashi, DR Congo', paper presented at the session 'The Multitude of "In-Between" Combinations in African Urban Spaces', at the EAUH's 12th International Conference, Lisbon, September 2014.

Fig. 1.3 An example of the signage system of Mutengene, as inscribed by hand by authorities of Société Nationale de l'Electricité du Cameroun (SONEL), the quasi-national electricity corporation in Cameroon (Editor's photo)

methodologies. But the main argument here is that enhanced methodological awareness is *traditionally* inherent in toponymic research that deals with the African continent, independently of the urban variable. This leads to a wide range of data and to an accumulation of rich research experience to inspire place name scholarship well beyond the continent. We hope that the present collection vividly contributes to such an inspiration.

1.2 Colonial Urban Legacies, Entangled Histories

The urban component matters. Though the urban population of Africa is still below 50 %, most of its sub-regions are highly urbanised (Northern and Southern Africa) or are most rapidly urbanised (Western, Eastern and Central Africa). The rapid growth of cities and towns in Africa has generally resulted in urban slum proliferation – this is due to a variety of reasons, amongst them inadequate urban governance, high levels of socio-economic inequality and segregation. Yet the failure to embrace imported development models by newly independent African states is now well understood, as is the need for major and creative rethinking of urban transitions, in a way that faithfully reflects actual conditions in situ.[46] One of these conditions in an era of globalisation is the current toponymic ambiguity or non-existence of formal naming systems in many sub-Saharan African cities. This directly impacts on their ability to benefit from cutting edge technology such as GPS, mobile telecommunications, ambulance response and other location-based services and business opportunities that necessitate precise toponymic property identification. Aside from some innovative remedies that have recently been described,[47] this situation also bears some symbolic implications in urban postcolonial Africa; for example, the very possession of a unique address is being perceived as a statement of high social status.[48]

But it is surprising how little research is available on the grounding of such informative and symbolic dimensions of African toponymy deep in the colonial past – where another key component, that of colonialism, also matters. In the colonial city, the 'nerve centre' of colonial economic exploitation,[49] street and place naming in the occidental sense were not only physically confined to the very limited space of the European quarters. The spatiality of colonial nomenclature was also a consequence of residential segregation on racial grounds, which simultaneously contributed to the constitution of residential segregation and further marginalisation of the 'Other.' This dual urban system, promoted de facto by the regimes of all European powers in Africa in spite of their stated variance with colonial ideology, referred only to the European quarters where western standards were applied in terms of urban layout, services, building regulations, commemorative street names and numbering.[50]

[46]UN-HABITAT, *The State of African Cities 2014: Re-Imagining Sustainable Urban Transitions* (Nairobi: United Nations Human Settlements Programme, 2014), esp. pp. 14–59; Jennifer Robinson, 'Global and World Cities: A View from Off the Map', *International Journal of Urban and Regional Research*, 26, 3 (2002), pp. 531–554.

[47]Ambe Njoh, *Planning in Contemporary Africa: The State, Town Planning and Society in Cameroon* (Aldershot: Ashgate, 2003), pp. 335–260; Ambe Njoh and Liora Bigon, 'Adapting Modern ICTs to the Spatial and Cultural Environment of Urban Africa with Emphasis on Cameroon', presented at the 6th Annual International Conference on ICT for Africa, Yaoundé, Cameroon, 1–4 October 2014.

[48]Coetzee and Cooper, 'The Value of Addresses to the Economy.'

[49]Anthony D. King, *Urbanism, Colonialism and the World Economy* (London: Routledge, 1990), p. 14. See also: Terence McGee, *The South East African City* (London: Bell and Sons, 1967), pp. 56–57.

[50]As for British colonial urbanism, King's 1976 study on British India was innovative in this respect and is certainly still unique. By dwelling on key generic notions such as the 'cantonment', 'mall' and 'bungalow-compound complex', King illuminated the reliable connections between classifying terminological systems and the colonial space, conceived as a social, cultural, behavioural and perceptual space. Though King is thin on evidence regarding the indigenous conceptions of settlement organisation and terminologies, his ability to de-familiarise and revisit his own British culture was then remarkable. As far as we are aware, this was the first book-length study to deal extensively with the relationships between language, social organisation, and physical urban form: Anthony King, *Colonial Urban Development: Culture, Social Power and Environment* (London, Henley, Boston: Routledge, 1976). For less systematic generic terminologies of colonial

Fig. 1.4 'Boulevard de la République', Dakar, Senegal (Editor's photo)

Designed and designated for the benefit of the expatriate population, these quarters and their attached spatial terminology reflected the mentality and ambiance of the politically-dominant culture, introducing – in a variety of global colonial contexts – a sense of uncanny order into what had been perceived as an amorphous, if not savage, landscape. Moreover, in many cases the actual toponymic inscriptions in these quarters testify to the ruling philosophy of the respective colonial power. To mention two, almost arbitrary cases: 'Boulevard de la République' in Dakar's city centre exemplifies the centralist-assimilationist thought of the French regime, though such representation of republican values in the colonial context is somewhat ironic (Fig. 1.4). Under the policy of assimilation, urban toponymic representation of indigenous cultures was close to null. In contrast, 'Ajele Street' right in the city centre of Lagos, Nigeria, testifies to the inclusion of indigenous vocabulary within the official signage. This points to the inherent flexibility of the British 'indirect rule' approach, with the word 'Ajele', which means 'Consul' in Yoruba, referring to the original location of the British Consulate (Fig. 1.5).

---

urban landscapes in some other contexts see the 'classical' works of: Janet Abu-Lughod, *Rabat: Urban Apartheid in Morocco* (Princeton: Princeton University Press, 1980); Alain Sinou, *Comptoirs et villes coloniales du Sénégal: Saint-Louis, Gorée, Dakar* (Paris: Éditions Karthala, Orstom, 1993); Gwendolyn Wright, *The Politics of Design in French Colonial Urbanism* (Chicago, London: The University of Chicago Press, 1991); Brenda Yeoh, *Contesting Space: Power Relations and the Urban Built Environment in Colonial Singapore* (Oxford: Oxford University Press, 1996).

Fig. 1.5 'Ajele Street', Lagos, Nigeria (Editor's photo)

In contrast, and the very existence of the contrast played a vital role in the formulation of colonial imagery, the spatial structure of the indigenous quarters was non-descriptive. But this top-down 'non-descriptiveness' was different from the bottom-up indigenous toponymic traditions, which were also generally non-descriptive and non-commemorative (the category of 'proper names' is limited strictly to Arabic names, Berber and Moor): these were either 'natural' names that related to local environmental features, geography, fauna and animals; or were connected to the historical or legendary origins of the setting, ethnic group, religion or other practices.[51] In the colonial context, most strikingly, the indigenous town, the one that gradually grew just beyond of the colonial administrative, business and residential 'centre', does not normally appear as part of the colonial urban plans at all. Its peripheral position was not only a quantitative physical matter, related to the general neglect on the part of the colonial authorities for the provision of basic infrastructure and amenities – an approach towards urban informality that was borrowed, together with its accompanied planning legislation, by the post-colonial governments.[52] It also represented conceptual neglect.

[51] Louis-Ferdinand Flutre, *Pour une étude de la toponymie de l'AOF* (Dakar: Université de Dakar, Faculté des lettres, 1957).

[52] See, for instance, the two-volumes collective project edited by Robert Home: *Essays in African Land Law*; and *Local Case Studies in African Land Law* (Cape Town: Pretoria University Press, 2011). And: Nancy Odendaal, 'Rethinking Urban Planning in the African Urban Century', a lecture presented at the International IPHS Conference 'Colonial and Postcolonial Urban Planning in Africa', Lisbon, 5–6 September 2013.

Being excluded from (post-)colonial city maps, the informal settlements, which today sometimes encompass up to 70 % or more of the city, have yet to be regarded as an integral part of the urban conglomeration, but rather as an object apart from the world. That is, the spatially-limited and exclusive model-space of the *ville officielle* (which in the colonial generic jargon is the *Plateau*, *ville nouvelle*, Hill Station, cantonment, Garden City, *ville européenne*, *ville blanche*, or *zone urbaine*) has been binary and essentially structured vis-à-vis the *ville réelle* (namely, the colonial *ville africaine*, *cité*, native village, *village de noirs*, *quartier indigène*, the traditional city, semi-urban zone, *bidonville*, etc.). In fact the problem is multi-layered. As expanded by Catherine Coquery-Vidrovitch, in terms of historiography, western urbanists were preoccupied with 'rural Africa', and were convinced that the (modern, colonial, contemporary) city is an innovation in Africa. "The reason for this", she says, "is that, consciously or not, but always implicitly, there is a presupposition that the only fully developed urban model is the Western standard of reference."[53] Similarly, in the official mind, to borrow Achille Mbembe's wording for our purpose, urban informality and traditional settlement forms still generally represent "the intractable, the mute, the abject, a failed and incomplete example of something else."[54]

Yet the aforementioned examples constitute only a glimpse. Possibilities for toponymic conservations regarding both past and present with a strong affinity to the colonial period are myriad. This is skilfully demonstrated by this volume's contributors, who refer to a variety of eras, urban perceptions and practices, and planning cultures. Therefore, the word 'legacies' in the current sub-heading not only points to the multiplicity of colonial regimes that were involved in the creation of colonial urban spaces and their generative toponymic traditions –our contributions deal with the British, French, Belgian and Portuguese heritages. The word 'legacies' in this collection equally implies the continuity of the colonial past and the reciprocal relations between colonial urban spaces and both the pre-colonial and the post-colonial periods in terms of place names. The schematic outline below exemplifies this relational approach:

[53]Catherine Coquery-Vidrovitch, *The History of African Cities South of the Sahara: From the Origins to Colonization*, trans. by Mary Baker (Princeton: Markus Wiener Publishers, 2005), pp. 12–13.

[54]Achille Mbembe, *On the postcolony* (Berkeley: University of California Press, 2001), pp. 1–4.

THE MULTIPLICITY OF INTERACTIONS AND DIFFUSION CHANNELS OF ENTANGLED TOPONYMIC PRACTICES IN URBAN AFRICA (CASE-STUDY-SPECIFIC/COMPARATIVE RESEARCH)

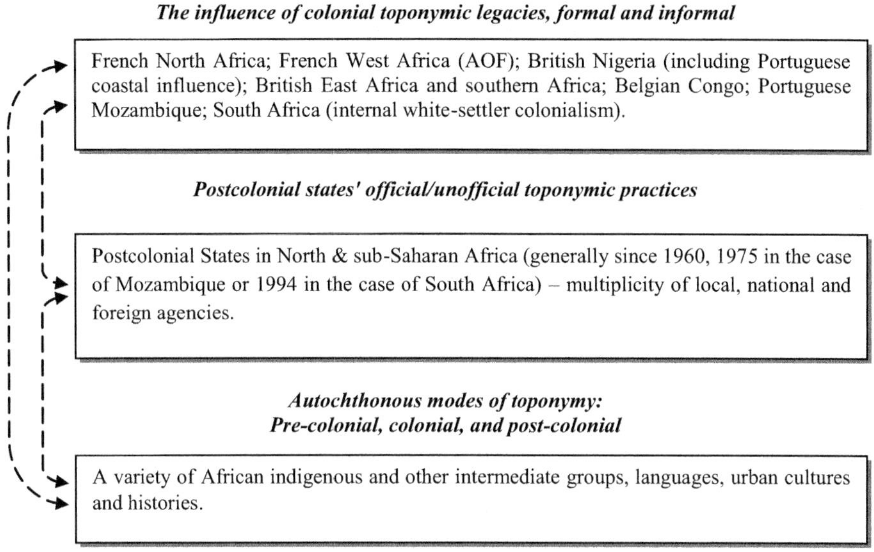

We found the relational approach – expressed in the term 'entangled' in the title of this volume – more suitable for this team project than a comparative approach per se. Some limitations of the comparative approach are derived from its external vantage point, which ideally strives to generate a symmetrical view of the objects in question; the need for equivalent observation in space, time and scale (micro/macro-histories); and consequently, the frequent conflict between synchronic and diachronic logics. Further difficulty may arise when the objects investigated for comparison display interaction. This is further complicated, as in our collection, in cases of multilateral comparisons that involve a large number of parameters.[55] The term 'entangled' is therefore highly suggestive for the nature of this collection. It offers a wide range of inter-crossed trajectories that are inherently open for inclusion, fragmentation, unevenness, hybridisation, creativity and a level of abstractness and imagination concerning spaces, points of time, scales, and cultural and socio-political experiences.[56] The relational approach of

[55] As elaborated by Michael Werner and Bénédicte Zimmermann, 'Beyond Comparison: *Histoire Croisée* and the Challenge of Reflexivity', *History and Theory*, 45, 1 (2006), pp. 30–50.

[56] Because the perspective of 'entangled histories' is somewhat undefined, thereby promoting creative flexibility and indefinite conclusions – it was also preferred here to other relational but more specific, approaches. Amongst the latter is *histoire croisée*, which developed as an explicit methodological 'toolbox' in order to produce particular knowledge effects (Werner and Zimmermann, 'Beyond Comparison'); or 'transnationalism', which essentially undermines or bypasses the 'national' factor by expanding on alternative channels of dissemination – yet the

'entangled histories' also "overcomes both the methodological nationalism and Eurocentrism of the social sciences",[57] but here our focus on Africa does not aim to "provincialise Europe"[58] or to marginalise the studied region by subjecting it to 'Area Studies.' 'Entangled histories' elaborate the grammar of toponymic diversity across local, regional, national and transnational spatialities and urban environments. This perspective also interrogates a neat demarcation of time periods and dichotomies between the micro and macro levels, enabling us to trace through the thematic leitmotif of place names, the capillary flows of power, memories and urban cultures.

1.3 Book Structure and Chapters Outline

Because of the high level of interference amongst the contributions between the local and regional (micro-level) and the supra-regional, multilateral, transnational, international and the global (macro-level); and because most of the contributions examine processes that took part in more than one time period (pre-colonial, colonial, post-colonial) – we found it both impossible and unnecessary to divide the book's chapters into parts according to theme, scale, geography or chronology. While we did not employ any synchronic, diachronic, spatial or analytical division of chapters in order to group the 15 contributions into several typological parts, a 'soft' chapter arrangement is still informally employed. This will enable the reader to gradually circulate between time spans, sub-regions and themes, though not in a linear-teleological way. This chapter arrangement invites possible inter-crossings and entanglements on a varied and multi-scopic basis.

Following this introductory chapter, Chap. 2 addresses the not always complementary and harmonious encounter between indigenous and exogenous modes of spatial perception and naming of (urban) settlements in Africa. Globes, maps, and gazetteers, notes Michel Ben Arrous, are not typically familiar with mobile spaces and places that have no fixed coordinates or precise limits. This chapter focuses on the mobility of Africa's social and spatial formations past and present, and provides a conceptual critique of the normative equivalence between geographic fixity

'national' still remains, even subconsciously. This complicates references to the pre-colonial period and even to the post-colonial present in Africa.

[57]Shalini Randeria, 'Entangled Histories of Uneven Modernities: Civil Society, Castle Solidarity and Legal Pluralism in Post-Colonial India', in Heinz-Gerhard Haupt and Jürgen Kocka (eds), *Comparative and Transnational Histories* (Oxford, NY: Berghahn Books, 2009), pp. 77–105 (p. 80). Randeria's conceptualisation of the term 'entangled histories' is inspiring for this collection though she focuses on historical and contemporary entanglements of multiple modernities between western and non-western societies. In this collection the meaning of the term was further diverted in order to meet its aims, structure and nature as explained in this paragraph.

[58]Dipesh Chakrabarty, *Provincializing Europe* (Princeton, Oxford: Princeton University Press, 2000).

and toponymic recognition. Building on examples from North, West, and Central Africa, Ben Arrous points out that African societies are not always where linear boundaries and conventional toponyms say they are. Common assumptions about the desirability and feasibility of toponymic standardisation are also questioned.

As a more specific case-study example, the histories and historiographies of the toponym 'Bagamoyo' are explored in Chap. 3 by Walter Brown (1943–2006). This chapter shows that during the late eighteenth century and early nineteenth century, a site connected regions and communities of a variety of languages and cultures in East Africa through long-distance trade. From a methodological point of view, this short but outstanding study is an excellent example of the 'traditional' problematic faced by Africanist historians. In trying to understand the past, these historians usually have to compare critically oral and written sources of rich linguistic origins – local, regional and foreign. They also have to navigate between several disciplines, as mentioned above, which is skilfully performed by the author. This is further complicated by the existence of at least 11 other places called 'Bagamoyo' on the Tanzanian mainland and environs. In fact Brown's contribution is reminiscent of a sentence that appeared on a leaflet for visitors distributed in the East African coastal town of Gedi, stating that "[a]n old town is like an old book, with some pages missing or torn, and some stuck together. The missing pages are lost forever, but the torn pages can be reconstructed and the stuck pages reopened."[59]

Also touching this geographic sphere and stressing the importance of bottom-up toponymic inquiry, Chap. 4 argues for taking the indigenous cultural practice seriously in developing urban environmental ideas about toponymy in Africa's cities. With a focus on Lusaka and Zanzibar, Garth Myers digs into the contrast, or apparent contrast, between the two cities in urban toponymy related to trees and neighbourhood names. This toponymy is tested against the realities of the actual socio-environments of these two cities in the contemporary context. Conceptually, Myers builds from a 'new' cultural-geography framework developed by Denis Cosgrove and interpretive methods developed by James Duncan to extend the analysis of intersections between cultures and environments in urban place-names and cultural practices-in-place. Myers argues that if we are to fully read urban environments on the continent and their toponyms, we need to have a more nuanced framework than one which poses power versus counter-power as a mere duality in urban politics. Everyday cultural conceptions of urban environments and place-names are therefore crucial, often hidden dimensions of the 'produced ecosystems' in Africa; and they offer a valuable lens on why a diversified and nuanced form of urban political ecology is needed for approaching the continent's cities.

In the same vein of acknowledging the importance, consistency and continuity of past toponymic conceptions well into the present in African societies and geographies, Chap. 5 takes us to West Africa. It may seem curious, argue Liora Bigon and Robert Home, that Lagos, a city of many millions, recognised as a world metropolis, should still be known by most of its Nigerian inhabitants among

[59]East Africa Tourist Travel Association, *Gedi: Royal National Park* (London: n.p., the 1980s).

themselves as *Eko*. This chapter aims to trace the toponymic history of Lagos over the 400 years since its original settlement, making analytical connections between place naming and built form. From Oko to Eko, from Curamo to Onim, and from Onim to Lagos – each of these name transfers represents a layer in the city's rich and cosmopolitan past. Through multiple primary and secondary sources (written and oral histories, cartography and architectural evidence), the shifting meaning of toponyms within the urban complex is scrutinised. Transnational and multilateral aspects of sub-Saharan Africa's history are highlighted, with the aim of exposing the complexities of simultaneous usages in Lagos's names. These constitute a fertile ground for juxtaposed memories on the part of the agencies, ethnic groups, and political, economic and cultural powers involved – memories which sometimes challenge one another, and are sometimes complementary.

The subsequent contribution that deals with North Africa's urban toponymy from a long-term perspective, analyses Fes in Morocco. Stressing place naming discontinuities of the French colonial regime, it also points to the vivid toponymic expressions of nationhood in post-independence time. Though most dominant in the public sphere of the present, this Arabised aspect of socio-political indigeneity is not the only one reminiscent of the past – some relics from antiquity have applied their toponymic seal to the urban memory as well. In Chap. 6 therefore, Samira Hassa uses four maps of the city of Fes from the years 1908, 1933, 1953 and 1986. Through these maps, the evolution of the toponymy of Fes, a post-colonial city, is examined across three historical eras: the pre-colonial period, the French occupation, and 30 years after the end of the French protectorate. Similarly to several other main urban centres that experienced French colonisation in North Africa, Fes is split into two well-marked districts: the Ville Nouvelle or 'New City', built by the colonial authorities; and the Medina, the old town, with its Muslim traditions and Arab urban architecture. While the toponymy of the Medina did not change, the Ville Nouvelle underwent changes in street naming that reflect changing political and social conditions. For instance, the 1933 map shows many street names honouring French military heroes, while the 1953 map shows more street names commemorating French intellectuals and an increase in Arabic street names, suggesting that the colonial French power was fading. In contrast, the 1986 map indicates that the toponymy gradually switched from French to Arabic in the Ville Nouvelle as result of the Moroccan nationalist movement. These changes of denomination, as noted by Hassa, are a symbolic reclamation of the previously colonised space and an affirmation of Moroccan-Muslim identity.

Remaining in the francophone zone, but going southwards to sub-Saharan Africa, Hélène d'Almeida-Topor investigates, in a comparative context, the exogenous colonial naming system of the capital cities of French West Africa. Building on official sources, Chap. 7 classifies this street naming system that was carried out by an official act, into several prominent themes. The reoccurrence of these themes in different rates in each of the capitals and the changing naming preferences along the years of colonial rule – e.g. the creation of new toponymic categories and the fading of others – testify to the character of the French regime in the region. Such toponymic classification also reflects an internal hierarchy within the vast territories of French West Africa in terms of political status and nuances of colonial imagery. A glimpse of the somewhat ambiguous re-appropriation of the French naming model in post-independence times is also offered by d'Almeida-Topor.

The comparative perspective is embraced in Chap. 8 by Luce Beeckmans as well, focusing on the colonial segregationist toponymy of the 'neutral zones.' The latter 'buffer' zones between the European minority and the African majority were popular in the colonial cities of sub-Saharan Africa, no matter which imperial regime. Expanding on British Dar es Salaam, French Dakar and Belgian Kinshasa, it is shown that 'neutral zones' toponymy was used as an apparatus for legitimising spatial segregation on a racial basis in these cities. In addition, the connection between this toponymy and transnational networks between the colonial forces of the region is highlighted, including a short overview regarding the fading of 'neutral zones' terminology in the era before and after de-colonisation.

In its further analysis of colonial segregationist toponymies, Chap. 9 by Liora Bigon and Yossi Katz examines the various planning practices that stand behind the generic term 'garden cities.' This is through an analysis of the inclusiveness and flexibility that are rooted in the term. While garden cities rhetoric and modes of planning in the western world are well covered in research literature, their counterparts in the southern hemisphere or in colonial contexts have gained relatively little attention. In scholarly works, garden cities notions and practices in colonial contexts have been considered simplistic at best, mere distortions of the original British models. The chapter traces and expands on the dissemination of garden city ideas in the early twentieth century from Britain to French Senegal and Ottoman Palestine. By bringing together Dakar and Tel Aviv (Ahuzat Bayit) our aim is not only to contribute to garden city historiography by an in-depth consideration of 'other' geographies. The aim is equally to acknowledge the inherent dynamism that is rooted in both garden city terminology and implementation, which corresponds to a rich variety of vernacular contexts (including the global South-East). Semantically and practically, these contexts therefore constitute an integral and essential part of the global history of the garden city planning phenomenon.

An extreme case of (internal) colonisation and accompanying racial segregation is pre-1994 South Africa, and here in Chap. 10, Gordon Pirie excavates the place-naming process of Johannesburg's most renowned black urban area, Soweto. Through the details of how he traced the sequence of naming proposals and their distinctive meanings, the discussions around the proposals and selection procedures, Pirie introduces us to the socio-political atmosphere of the period. The specific details of this case are telling examples of the contemporary limitations, agencies, possibilities and operational channels that were involved in such a top-down colonialist act of naming.

The post-colonial element becomes more dominant in the following four contributions, of which Chap. 11 discusses the post-apartheid toponymic landscape of South Africa. This landscape, argues Barbara Meiring, is as colourful as South Africa's flag, a kaleidoscope of colour that reflects the diversity of South African citizens. Meiring aims to point out that the geographical names, in this case of urban entities, also collectively reflect the hearts and minds of those who gave the names, ultimately expressing what it means to be a South African in the face of shared historical and cultural experiences behind the allocation of the name. The effect and complexities of the post-1994 transformation policy is not ignored, focusing on how transformation differs from mere change and how these two concepts apply to the geographical names of South Africa and its peoples. Using the onomastic framework of the etymological and associative meanings of place names, this chapter points out that being South African involves both shared and differentiating experiences, leading to the conclusion that there is no definition of a South African identity and that a national identity is subject to a consciousness and tolerance of social diversity and need.

Chapter 12, written by Johan Lagae, Sofie Boonen and Donatien Dia Mwembu Dibwe, expands on several toponymic episodes in the urban landscape of the Congolese city of Lubumbashi through the ages. The changing names of the selected streets and avenues from the colonial period till the present illustrate the socio-political power of toponymic inscription under changing regimes in shaping the everyday memories and counter-memories of the urban residents. But some renaming endeavours can be defined, in the examples provided, as impulsive rather than thoughtful acts, a consequence of post-colonial amnesia rather than of any historical awareness. Drawing on archival research, fieldwork and local history, the authors argue that the changing toponymy of Lubumbashi offers a particular insight in the shifting position of the city, and, by extension, the province of Katanga, in the minds of local as well as central authorities in Congo. One of the outstanding features of this chapter is its unique contribution to the study of multi-layered renaming processes. Mobutu's *'Recours à l'authenticité'* strategy had a twofold aim: erasing the colonial legacy as well as restoring national unity and common identity. Yet a tendency to re-introduce some colonial street names for cultural-historicist reasons has been recently noted in the capital, Kinshasa. Against the background of the separatist experience of Lubumbashi, the authors show, that recourse to a colonial name was useful in suppressing regional nationalism. Moreover, a name could now be detached from its typical Belgian colonialist connotation, now signifying indigenous memoires.

Embracing a socio-linguistic and anthropological approach to examine toponymic inscriptions, in Chap. 13 César Cumbe analyses the discursive functioning of formal and informal place naming in formerly Portuguese Maputo. As contextually-produced linguistic and social practices, official and unofficial place-naming systems testify to the historical and everyday lives of the urban residents, from both a top-down and bottom-up perspectives. Cumbe draws our attention to the fact that the simultaneous existence of official and unofficial toponymy in Maputo does not seek to replace or compete with each other. In the contrary, this situation only consolidates and enhances contemporary history and memoires, together with city dwellers' reflection of everyday reality and spatiality. Moreover, while toponymic informality exemplifies Maputo's urban multilingualism, it also neutralise the geographical and administrative boundaries of the urban sphere and challenges the centre-periphery dichotomy. This challenge takes place, inter alia, through the social aspects of transport that circulates throughout the urban area. Such transport exhibits a linguistic landscape that juxtaposes a variety of place names and popular shops, regardless of their actual socio-spatial location in the city. In this way, socio-economic and ethnic stratification of the city's various parts is virtually dissolved.

Chapter 14 is unique in its insertion of the international socio-political arena into the site-related nature of the street. Wale Adebanwi not only focuses on local and national street naming practices but also on how local politics intersects with international politics. This chapter examines the politics of spatial inscription and the social reproduction of 'place' or 'space' on a street corner in New York City named after Kudirat Abiola, an assassinated woman activist in Nigeria, and the retaliatory renaming by the military regime of a Lagos street housing the US Embassy, after the African-American anti-establishment activist, Louis Farrakhan. Later, the next democratic government of Nigeria renamed the street, this time after the US ambassador, the African-American Walter Carrington. Toponymy, the chapter concludes, can thus be seen as a form of retortion in international relations. Indeed, such a vantage point on the post-colonial present of Lagos in the international context is refreshing and casts a new light on the historicised Chap. 5, in which transnational aspects of Lagos' pre-colonial and colonial toponymy were discussed. Finally, our team effort of elaborating on the grammar of toponymic expressions in urban Africa is concluded in the Afterword, calling, inter alia, for further works focusing on the global South. This is in order to enrich our toponymic insights and historiographies by closing the gap between the northern and the southern hemispheres.

Chapter 2
Sarah's Globe and the (Un-)naming of Mobile Space

Michel Ben Arrous

Abstract Globes, maps, and gazetteers are not typically familiar with mobile spaces and places that have no fixed coordinates or precise limits. This chapter focuses on the mobility of Africa's social *and spatial* formations, past and present, and provides a conceptual critique of the normative equivalence between geographic fixity and toponymic recognition. Building on examples from North, West, and Central Africa, the chapter points out that African societies are not always where linear boundaries and conventional toponyms say they are. Common assumptions about the desirability and feasibility of toponymic standardisation are questioned.

Keywords Mobile spaces • Unconventional toponyms • Toponymic standardisation • Africa's geography • (Un-)official boundaries

> Colonisers prospered throughout centuries from naming what they appropriated, and unnaming what hindered them.[1]
> The Earth is a palimpsest, a graveyard of toponyms.[2]

Sarah, my 10-year-old niece, loves reading maps and learning new place names. She has a Moroccan mother, a Cape Verdean father, and they all live in Morocco. A globe, I thought, would make a perfect gift for her. So I got one and handed it to her father, who one day was passing through Dakar, my hometown. Things did not go easily. Sarah's father was stopped upon arrival at Casablanca Airport. There he was detained for several hours, not knowing why. Customs officers seized the globe, which he did not understand either. How could they not see that it contained neither drugs nor anything illegal?

Asked to come back the following day, he was finally informed that the globe could not enter the kingdom and had to be destroyed. There was nothing wrong

[1] Kamel Daoud, *Meursault, contre-enquête* (Arles: Actes Sud, 2014), pp. 22–23.

[2] Christian Jacob, *L'empire des cartes: Approche théorique de la cartographie à travers l'histoire* (Paris: Albin Michel, 1992), p. 309.

M. Ben Arrous (✉)
Dakar, Senegal, and Bordeaux, France
e-mail: benarrous@arc.sn

L. Bigon (eds.), *Place Names in Africa*, DOI 10.1007/978-3-319-32485-2_2

inside it, but on it. The object of the offence was a yellow area bearing the toponym Western Sahara, just below the green area with the name Morocco. A dashed line, the standard cartographic convention for disputed borders, ran between them. From a Moroccan point of view, the yellow area belonged to the Southern Provinces of the green one and any other mentions or allusions, including toponyms, were banned. It did not really matter that most people outside Morocco refer to the disputed area as Western Sahara, its official UN name (see Fig. 2.1: map features the toponyms that are mentioned in this chapter). A 'non autonomous territory' in UN parlance, the former Spanish Sahara is a full-fledged member of the African Union, where it is known by the acronym DASR, or Democratic Arab Sahrawi Republic. Had the globe makers printed the latter name, things could have gone much worse for both Sarah's gift and her father.

Western Sahara was the only unacceptable toponym. Fortunately, a sympathetic customs officer had the bright idea of erasing it with a razor blade. This was how, just as in Daoud's first quote above, the hindrance to pleasing my niece was overcome: simply by unnaming it. So name suppression proved useful and could not be considered, contrary to what Daoud seems to imply, a colonial-specific operation.

Further erasure of the toponymic palimpsest would have revealed several layers of Spanish, Arabic, Hebraic, and Amazigh names – some of them dead, some not. A toponym may have many lives, depending on the purposes to which it can be put. For instance, the Western Sahara's independence-seeking 'Popular Front for the Liberation of Saguía el-Hamra and Río de Oro' (POLISARIO) chose to name itself after the two 'Spanish territories' that once made the Spanish Sahara. The reuse of colonial toponyms here establishes an explicit frame of reference for decolonisation claims. On the other hand, the positional adjective 'Southern' for these territories in Morocco's Southern Provinces involves some larger, encompassing space with a political centre of gravity further north. But the very name of Morocco (or its most common one, besides the Arabic al-Mamlakah al-Maghribiyyah, i.e. 'Kingdom of the West') derives from the Portuguese pronunciation (Marrocos) of the Almoravid-founded city of Marrakech, or Murwkush.[3] The Almoravids, a confederation of Sahrawi tribes from the area of today's Mauritania, ruled over an empire that in the eleventh century stretched from central Spain to the shores of the River Senegal and included the whole of present-day Morocco.

[3]Marrakech is an Amazigh toponym with uncertain meaning(s). Allegedly based on the verbal root RKS ('to hide'), derogatory meanings like 'the place where [hidden] bandits attack caravans' or 'decamp quickly [and hide]' were quite popular in colonial times. Current linguists rather focus on the Amazigh words *mur* ('rescue' or 'protection'), *amur* ('protected place'), *Kush* (grandson of Biblical Noah) or *akush* ('divinity') to reinterpret Marrakech as 'land of God', 'mountain of Kush', or a 'blessed place [. . .] where respect of a divine pact excludes violence.' See A. Toufik, 'Marrakech: La signification du nom', in Salem Chaker (ed.), *Encyclopédie Berbère* (Louvain: Peeters, 2010), vol. 30, pp. 4627–30.

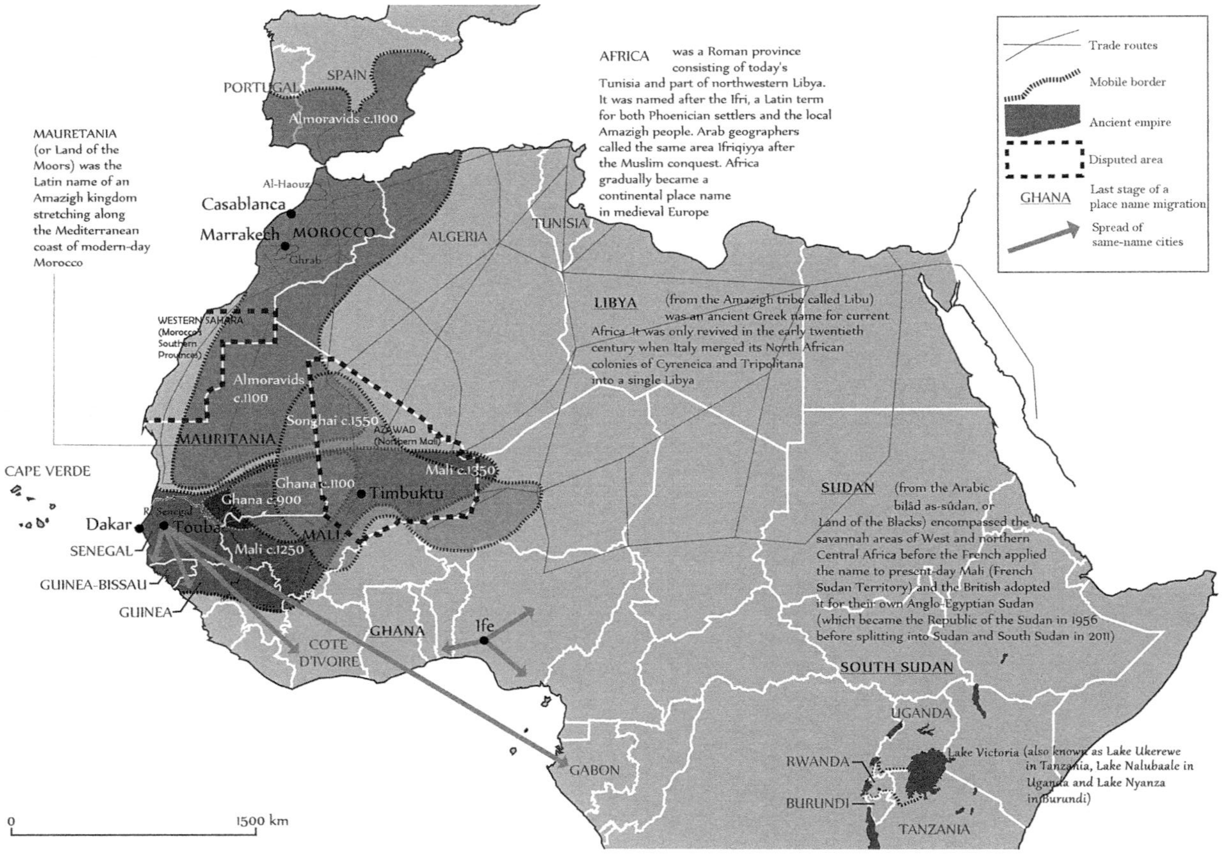

Fig. 2.1 Map features the toponyms that are mentioned in this chapter

In short, the green area on Sarah's globe stands for a country whose name somehow stems from the remote ancestors of those who, in the yellow area, ask that the dashed line between green and yellow be made again as solid as when it marked an inter-colonial boundary between two other colours. That is, beige-tinted Spain (half of which they once ruled), and France in light purple (who held the green area as a protectorate from 1912 to 1955). If a scratched globe says anything about names and borders, it is their shared symbolic importance. Names and borders have become, at all latitudes, key attributes of a territory. The most enduring colonial legacy in this regard, the one with the most far reaching implications, is the delineation of continuous spaces as a prerequisite to naming.

The scraped off yellow area on the globe was never a territory unified under a single authority. It was part of different linguistic, social, and economic spaces that varied in extent and overlapped. Its structuring (physical) elements such as trade routes, market places and stopping points, shaped adaptive networks with changing configurations and varying levels of control by various groups in various alliances. Even when prominent tribes extended *beya* (allegiance) to the Sultan of Morocco, or also to the Dey of Algiers or to Songhai rulers to the south, the remaining space contained the largest area and was commonly known as *bled es-Siba*, or 'land of dissidence.' No erasure of whatever globe could reveal this generic *bled es-Siba*, for the simple reason that it refers to an ever-changing space with no fixed limits. Globes, maps, and more generally our ways of thinking and representing space have become unfamiliar with this kind of geography.

On the mural map in her classroom, Sarah has learnt to locate every subdivision of Morocco. When asked for the Al-Haouz Province for instance, or for the Ghrab Region, she identifies the relevant administrative boundary first, and then she follows it with her index finger in a circular movement. She is not expected to know that *gharb* ('west', as in *maghrib*) and *hawz* ('environs', 'vicinity') had long been a very broad division of the Atlantic plains between coastal areas and their hinterlands, nor that *Hawz* later came to refer to the hinterland of every city of importance. This imprecision remained unproblematic at least until the late sixteenth century, when the Council of Portugal, who ruled over most of the Atlantic coast and harbours, began pressing for toponymic disambiguation.

Delineation and disambiguation enabled colonial inventories of the world. They contributed to the cartographic, then political, closure of hitherto open spaces, where most African societies, not only nomadic ones, used to expand, retreat, or simply move depending on political, economic and environmental conditions and circumstances.

The notion of fixed borders was alien to pre-colonial entities. Even those which had linear boundaries, such as the Great Lakes kingdoms in Central Africa, conceived them as intrinsically mobile. Their movement in Rwanda was codified through a dynastic cycle of four kings, two of whom had a duty to conquer new lands, while the other two were confined to 'sacred places' and forbidden to wage war – their responsibility was to restore peace and consolidate the enlarged kingdom. Consequently, a same hill or village could receive different names at different moments in the dynastic cycle, according to its spatial and symbolic integration within fluid borderlands.[4]

Mobile spaces and changing toponyms may be more relevant to the understanding of Africa's geography, past and present, than the classical distinction between nomadic and sedentary peoples or the cartographic and toponymic grids against which we are educated to report movements. The seasonal movements of those we unthinkingly call 'nomads', in and around the Sahara or elsewhere, actually take them to more or less the same places year after year. Their social and material practices thus produce quite stable spaces. On the other hand, many of the peasant societies we consider 'sedentary' have been living off itinerant agriculture for centuries and are indeed much more 'geographically' mobile. Acknowledging not only physical migrations from one named place to another named place, but the spatial mobility of identities and cultures, including the dynamic reconfigurations of linguistic, religious and ethnic areas, is basically a matter of time scales. African societies, in other words, are not always where official boundaries and affixed toponyms say they are.

*

To name a place is to differentiate it, to single it out from a broader spatial continuum. Nowadays this involves the mental mapping of discrete geographical entities – a cognitive operation that we perform without effort, anytime we mention a place or one is mentioned to us. Even in cases of name dispute, as Southern Provinces vs. Western Sahara on Sarah's globe (or the Falklands vs. Malvinas, or Judea and Samaria vs. Jordan's West Bank, etc.), we tend to focus attention on social, political or cultural narratives and significance, i.e. on conflicting constructions of placeness, but one name or the other evokes for all parties a single geographic location.

The meanings we attach to places are by contrast too volatile to be captured, let alone stabilised, in a single toponym. Meanings evolve as intricate outcomes of events big and small, memory constructions and contextual subjectivities, while toponyms rarely change outside major political shifts. Place names provide historians with benchmarks regarding the name givers, the language they spoke, the social and power relations that shaped their times – but they do not say much about what the named places mean to whom. Rather, toponyms serve as supports of, and stimuli for, geographical imaginations, allowing different people to invest the same portion of Earth with different meanings – thus making it different places at the same time.

[4]Christopher C. Taylor, 'Fluids and Fractals in Rwanda: Order and Chaos', in Mark S. Mosko and Frederick H. Damon (eds), *On the Order of Chaos: Social Anthropology and the Science of Chaos* (New York, Berghahn Books, 2005), pp. 136–165.

Multiple names would seem a natural way of dealing with multiple meanings. They are widely regarded as anomalies. Multiple names are a nightmare for gazetteer editors, and remain last resort exceptions to the 'one place, one name' principle. Or, more exactly, to the established standard of 'one *specific* location, one *fixed* name.'

All named places, in other words, are implicitly understood as having stable geographic coordinates. If *nomen est numen* (to name is to know), as the Latin adage goes, then both familiar and exotic places, once given names, are knowable by their location on maps. Without even a vague idea of where they are, the simple fact of using toponyms indicates that they exist somewhere, as mappable entities. This eases interpersonal communication and other aspects of social life, ranging from mail delivery to land titling and tax recovery, not to forget GPS navigation. The reverse side is that places with imprecise or moving positions – and many such places are meaningful to many people – remain below the radar of toponymy.

Places are acknowledged as being places when they are named. And they are named when we can map them in a steady position. As tautological as it appears today, the normative equivalence between named and mappable places long remained unnecessary in most parts of the world. A byproduct of the European Renaissance, it gradually came about with the turbulent decline of the Holy Roman Empire, where popes and kings claimed a right over people, and the development of secular states that asserted a right over land.

The territorialisation of sovereignty, the development of land measuring techniques and survey methods, and the expansion of a Eurocentric mercantile world-system, converged towards a paradigm of geographic permanency. Space was made a supposedly inert canvas on which the movements of history could be daubed. Well beyond state borders, the endless search for stable divisions has affected all scales of geographic imagination, from our pseudo-evident five- to sevenfold continental system to babushka-like structures, at sub-state levels, of nested regional and local governance.[5] Ironically, the notion of fixed territories spread to Africa first as a colonial prescription, then as a condition for the possibility of decolonisation.

What were decolonised peoples supposed to do with colonial borders? To keep them, erase them, or renegotiate them on a case-by-case basis? Their official freezing, after heated debates, proved tantamount to renewing a basic form of coloniality[6] or a distinctively colonial way of shaping African social formations. New flags were raised, new anthems sung, and new names given to a number of countries, cities, streets, airports and other symbolic places. But who belongs to what society, who decides so, who participates in collective identity formation (and transformation), turned out to be burning and conflict-generating issues.

[5]Martin M. Lewis and Kären E. Wigen, *The Myth of Continents: A Critique of Metageography* (Berkeley, University of California Press, 1997); Achim von Oppen, 'Bounding villages: the enclosure of locality in Central Africa, 1890s to 1990s' (unpublished habilitation thesis, Humboldt-Universität, Berlin, 2003).

[6]Sabelo J. Ndlovu-Gatsheni, *Coloniality of Power in Postcolonial Africa: Myths of Decolonization* (Dakar: CODESRIA, 2013).

The problem was not so much where borders are placed. It was, and still is, how people conceive sameness and otherness (i.e. placeness and distance) and above all how they conceive limits. As separating walls, or bridges? As thick lines fitting an all-encompassing territory? As identity gradients running through overlapping spaces? As an immutable given or varying constructs?

*

Algerian writer Kateb Yacine[7] asked what the very name of his country could mean to non-coastal populations: '*Ldjazaïr*, what's that name? Ever seen a country named "the islands?"' Kateb's target was an "arrogant [. . .] brand of Arabo-Islamism" that placed all Algerians in a reductive identity through the Arabisation of not only French names, but of Amazigh toponyms too, which obscured or actually concealed their 'real and deeper African dimension.'[8]

Renaming the landscape, whatever language is being used, is an act of power and an intrinsically divisive exercise.[9] It proves all the more disruptive when a plurality of peoples speaking different languages have historically interacted without thinking of themselves as being members of clear-cut entities, and 'intermingled to such an extent that intermarriages, linguistic and cultural exchanges resulted in the emergence of [large clusters] of distinct but interconnected peoples.'[10] This observation was made in a Southern African context but it could also apply to North or West and indeed all parts of Africa where inter-societal relations have historically 'formed not a mosaic of more or less autonomous communities, but rather genuine "chains of societies."'[11]

Against this general background of constantly ongoing societal reconfigurations and mobile spaces, UNESCO's commitment to 'urgently achieve a system of norms and standards for African [. . .] toponyms'[12] would seem at least contradictory with the organisation's very mission to foster peace through culture. There have been

[7]Yacine was his first name and Kateb his family name. He insistently asked to be known as Kateb Yacine. Reversing the usual name order was, in his view, a way of mocking the colonial civil system.

[8]Kateb Yacine, 'C'est africain qu'il faut se dire' [reprint of a 1987 interview with Tassadit Yacine], in *Le Poète comme un boxeur: Entretiens 1958–1989* (Paris: Seuil, 1994), pp. 101–120 (pp. 101, 109).

[9]Maoz Azaryahu and Arnon Golan, '(Re)naming the Landscape: The Formation of the Hebrew Map of Israel 1949–1960', *Journal of Historical Geography* 27, 2 (2001), pp. 178–195.

[10]Denis-Constant Martin, *Sounding the Cape: Music, Identity and Politics in South Africa* (Somerset West: African Minds, 2013), p. 350.

[11]Martin, *Sounding the Cape*, p. 350.

[12]UNESCO, *Ethnonymes et toponymes africains*, Histoire générale de l'Afrique: Etudes et documents, 6 (Paris: UNESCO, 1984), p. 9.

many contentious attempts at stabilising place names since independence,[13] all of them grounded in questionable assumptions regarding the desirability and feasibility of toponymic standardisation.

One single place may have different names, such as 'Lake Victoria' (which originated in colonial times after the British queen), which Burundians know as 'Nyanza', Tanzanians as 'Ukerewe', and Ugandans as 'Nalubaale.' Different places may bear one same name, as the many Ife-s established throughout West and Central Africa by Yoruba migrants, or the many Touba-s – such as in Guinea, Guinea-Bissau, Côte d'Ivoire, and Gabon – named after the spiritual capital of the Mouride brotherhood, in central Senegal, from where migrant disciples spread the message of their Sufi order.[14] This is aside from the countless districts known as 'Little Touba', because of high Mouride 'visibility' in among other places Rome, Paris, New York and Hong Kong.

At-tūb ('sun-dried mud-brick') or *tūbā* ('felicity', or also the 'tree of paradise'[15]), the usually considered Arabic etymologies, allow Mouride and non-Mouride inhabitants of these various Touba-s to imagine themselves within or outside a larger and mobile Mouride space. In the same vein, the acceptance of multiple possible etymologies for Timbuktu – a 'hollow' in Songhai, in reference to the depression where the city stands; or, in Tamasheq, the 'place' (*tim*) or the 'wall' (*tin*) of either a 'small dune' (*buktu*) or of a woman named *Butu* – allows for its inclusion within different cultural spaces simultaneously, irrespective of the changing state of community relations.

With all-encompassing territorial reforms backed by international donors' support, toponymic debates and grassroots participation in local naming commissions are now becoming an increasingly accepted norm throughout Africa.[16] A supposedly-essential ingredient in democratic governance, they look very good but, if the places and territories to be named or renamed have been mapped in advance, they rarely achieve more than pouring old wine into new calabashes. Once again, the ways we think about space determines the places to be named. Even in cases when the reconfiguration of territorial units is left for discussion, as happened in Mali since the mid-1990s, an imposed criterion of spatial continuity limits the significance of the whole process.

[13]Atoma Batoma, 'African Ethnonyms and Toponyms: An Annotated Bibliography', *Electronic Journal of Africana Bibliography*, 10, 1 (2006), see http://ir.uiowa.edu/ejab/vol10/iss1/1/ (visited 28 December 2014).

[14]See, respectively: Eva Krapf-Askari, *Yoruba Towns and Cities* (Oxford: Oxford University Press, 1969), p. 8; Eric Ross, 'Marabout Republics Then and Now: configuring Muslim towns in Senegal', *Islam et sociétés au sud du Sahara*, 16 (2002), pp. 35–66.

[15]Eric Ross, 'Touba: A Spiritual Metropolis in the Modern World', *Canadian Journal of African Studies*, 29, 2 (1995), pp. 222–259 (pp. 223–224).

[16]Frédéric Giraud, Myriam Houssay-Holzschuch and Sylvain Guyot, 'Au nom des territoires! Enjeux géographiques de la toponymie', *L'Espace géographique*, 37, 2 (2008), pp. 97–105.

Mali is indeed a revealing illustration.[17] The basic form of spatial grouping prior to colonisation was the *diamana*, a networked space with hazy borders and varying centralities. French colonisation established a rigid territorial structure, where the smallest unit directly headed by a colonial officer was the *cercle* (the French word for 'circle', conveying the idea of circumscription). Each *cercle* was named after, and centered on, a fixed *chef-lieu* (literally a 'chief place') and consisted of several *cantons* (from the verb *cantonner*, meaning 'to confine' or 'to station') that fell under the delegated authority of colonial-appointed chiefs. The *cercle* system was maintained at independence, except that *cantons* came to be called *arrondissements* (same principle of 'encircling') and the colonial chiefdom gave way to local delegates of the ruling party. Hundreds of new local communities have been recently created through inter-village consultations. The overwhelming majority of them are named after their (democratically chosen) *chef-lieu*, which more than anything attests to 'the difficulty of devising alternatives' to the model of the steady-space-having-a-steady-centre.[18] Names of former *diamanas* come in second position, thus escaping the notion of fixed centralities but not that of fixed delineation.[19] It would seem in this regard that the Earth is less a graveyard of toponyms than of toponymic models, i.e. of ways of thinking and naming space.

To conclude, one final remark about Sarah and her Cape Verdean father. Like many people of Sephardic descent, my niece has a family name that is also a Portuguese toponym – a symbolic form of mobile space, one that Jewish families carried with them on their expulsion from Spain and Portugal 500 years ago. Sephardic 'toponymic patronyms', a reminder of past home places in the Iberian Peninsula, are now common across West Africa, principally in the former Portuguese colonies of Guinea-Bissau and Cape Verde, in southern Senegal and some coastal areas of the Gulf of Benin – without most bearers being aware of

[17]Now a landlocked country, Mali takes its name from the Mali Empire, one of the most powerful and most mobile polities in West African history. Ancient Mali at its height, in the fourteenth century, encompassed about half the current Malian territory, stretched across present-day Mauritania and Senegal to the Atlantic Ocean, and had its capital in present-day Guinea. Toponymic displacement is a pronounced phenomenon in Africa – the prime example being Ifriqiya/Africa itself (once a Roman province centered on modern Tunisia), along with, inter alia, Libya, Mauritania, Ghana and Sudan. As Lewis rightly points out, however, the 'migration of place names [...] taking on different meanings as they are translated and as basic geographic conceptualizations change' is not that surprising, since 'change is intrinsic to language itself.' Martin W. Lewis, 'The Migration of Place Names' (2011), www.geocurrents.info/historical-geography/the-migration-of-place-names-africa-libya-ethiopia-eritrea-and-sudan (visited 28 December 2014).

[18]Stéphanie Lima, 'L'émergence d'une toponymie plurielle au Mali', *L'espace politique*, 5, 2 (2008), http://espacepolitique.revues.org/1115 (visited 22 January 2015).

[19]Still regarding Mali, another great paradox was the NMLA's (National Movement for the Liberation of Azawad) proclamation of a breakaway (and unrecognised) state in April 2012. The Tamasheq word *azawad* translates to 'land of transhumance' and it basically refers to an open area, one whose fuzzy borders vary according to climatic conditions, community relations, and other factors of uncertainty. By contrast, the NMLA conception of Azawad as being the northern half of legal Mali derives from, reflects and even reproduces the very territorial model that has long caused Tuareg rebel groups to take arms.

any Jewish ancestry. Recent government decisions, in both Spain and Portugal, to facilitate the restoration of citizenship for the descendants of expelled Jews have led to the publication and publicising of lists of Sephardic names that can be used to apply for naturalisation.[20] These lists, whatever their impact on Jewish diasporic spaces, and their possible help to African candidates for a European nationality, constitute one more instance of varying geographies and toponymic displacement. Both Sarah's globe, an emblematic object of Iberian circumnavigations, and Sarah's name, a symbol of mobile spaces, call for a critical reassessment of our toponymic standards and formats.

[20]Compiled lists available on http://my.ynet.co.il/pic/news/nombres.pdf (visited 28 July 2015).

Chapter 3
Bagamoyo: Inquiry into an East African Place Name

Walter T. Brown

Abstract This exercise in the histories and historiographies of the toponym 'Bagamoyo' was conducted by the late American professor Walter T. Brown (1943–2006). It shows that during the late eighteenth century and early nineteenth century, the site connected regions and communities of a variety of languages and cultures in East Africa through long-distance trade. From a methodological point of view, this short but outstanding study is an example of the 'traditional' problematic faced by Africanist historians. In trying to understand the past, these historians usually have to compare critically oral and written sources of rich linguistic origins – local, regional and foreign. They also have to navigate between several disciplines, what is skilfully performed by the author. This is further complicated by the existence of at least 11 other places called 'Bagamoyo' on the Tanzanian mainland and environs. In fact Brown's contribution is reminiscent of a sentence that appeared on a leaflet for visitors distributed in the East African coastal town of Gedi, stating that "[a]n old town is like an old book, with some pages missing or torn, and some stuck together. The missing pages are lost forever, but the torn pages can be reconstructed and the stuck pages reopened."

Keywords Bagamoyo • East-African languages • Toponymy • Swahili coast • East African history

Place naming can symbolise modes of living, an ideology, a period of crisis or periods that favour change. This phenomenon is exemplified in the toponym 'Bagamoyo', referring to an East African coastal town. The case of Bagamoyo is illustrative of a century during which regions, ethnic groups, communities and languages were linked together through long-distance trade. This chapter expands on several etymologies of the word Bagamoyo, as designated by Swahili-speaking porters arriving there from the hinterland during the late eighteenth or early nineteenth century. An analysis of the origin of this word is complicated by the

W.T. Brown (deceased)
African History and International Studies, Ramapo College, Mahwah, NJ, USA

L. Bigon (eds.), *Place Names in Africa*, DOI 10.1007/978-3-319-32485-2_3

variety of languages possibly involved, and by the existence of at least 11 other places called 'Bagamoyo' on the Tanzanian mainland and environs.[1]

It would be presumptuous to suggest that one word could neatly and succinctly capsulise any historical era. Yet the historian can readily utilise place names such as Masada, Machupijchu and Culloden Moor to symbolise a life style, an ideology or a period of crisis and change. So too can the name 'Bagamoyo' be illustrative of a century during which long-distance trade linked regions and communities of many different cultures in East Africa. These commercial lines became dispersion routes for not only people and goods, but also architectural, cultural and religious concepts. Unfortunately, the traditions of the origin of this Indian-Ocean coastal town and its name are not extensive, but the little direct evidence that has survived illuminates the kind of place it was – and in many aspects still is. An introductory portraiture of this multi-cultural and multi-racial town can be gleaned from an exercise in the etymological history of the word 'Bagamoyo.'

Amongst the earliest written materials describing Bagamoyo, the correspondence and journals of the Holy Ghost Fathers, are especially pertinent. In the late 1860s, one priest wrote that Bagamoyo in Swahili meant "jusque dans le Coeur" [into the heart] – thus indicating that the town was so named because it served as the path into the interior or heart of Africa.[2] This interpretation was later repeated by Jerome Becker and Georges Revoil who noted that it signified "Coeur de l'Afrique" [Heart of Africa] – derived from *baga* or *bana*: interior, and *moyo*: heart.[3] The missionary Ameet Vyncke suggested a similar conclusion by the yoking of *paka* (*mpaka*): until, and *moyo*: heart.[4] A more frequent explanation is that the word is a version of *bwagamoyo*: translated variously as "Be quiet, my heart",[5] "Lay down the burden of

[1]This chapter, originally entitled 'East African place names: an etymological inquiry', was published in 1973 in the Journal *Africana Marburgensia* by the late American professor Walter T. Brown (1943–2006), a specialist in the cultural, economic and political history of the East African Coast.

[2]*Bulletin Général de la Congrégation du St. Esprit et de l'Ime Coeur de Marie*, vol. VI, f 416; *Les Missions Catholiques*, vol. II (1869), p. 3; and vol. XIV (1882), p. 194.

[3]Lucien Heudebert, *Vers les grands lacs de l'Afrique oriental: d'après les notes de l'explorateur Georges Revoil* (Paris: Alcide Picard Editeur, no date [circa. 1900], p. 62; Jerome Becker, *La Vie en Afrique* (Paris: Lebègue, 1887), vol. I, p. 45. A Portuguese map dated circa. 1600 contains the word *Baga* (interior) written over a portion of central Africa. Armando Cortesao e Avelino Teixera da Nota, *Portugaliae Monumenta Cartographica* (Lisboa: Instituto dos Arquivos Nacionais, 1960), vol. III, plate 362C.

[4]Ameet Vyncke, *Brieven Van Eenen Vlaamschen Missionaris in Midden-Afrika* (Brugge: Beyaert-Storie, 1888), 2 de Reeks, p. 138. An edited and annotated translation of this book will appear in: Walter T. Brown and Marcel B. Bervoets (eds), 'Ameet Vyncke: From Bagamoyo to Ujiji in 1883. Selections from his Letters', *Tanzania Notes and Records*, forthcoming [the Editor failed to identify the item].

[5]A. Leue, 'Bagamoyo', *Beiträge zur Kolonialpolitik und Kolonialwritschaft* (1900), p. 13.

your heart",[6] "Rest your soul (here)",[7] "Free the heart",[8] and "Rest the mind, throw off melancholy, be cheered."[9]

Three interpretations of these translations have been developed. The first theory holds that the town's name originated with the lament of slaves who knew that, although they were to be shipped to distant lands, their hearts would forever remain in their beloved homeland. A second theory maintains that the two words were spoken by an anonymous founder of the town who, having abandoned previous sites which had been plagued by epidemics, decided to settle on this section of the coast.[10] The third theory, which I consider the most feasible, contends that Bagamoyo was named by *wapagazi* (porters) who felt that they had successfully completed their journey from the interior (usually the Tabora area and westwards) (Figs. 3.1 and 3.2). They were in the habit of congratulating themselves – "now cease from worrying, cease from care and anxiety."[11] A young boy from Manyema (in the eastern Congo) named Kabuera who had 'worked' for an Arab as a porter recalled the contentment and happiness felt by the porters as they laid down their loads for the last time at Bagamoyo. Having arrived safely, he thought, "I am at the end of the world, I see only the ocean before me."[12] The third interpretation is also bolstered by a report concerning caravan porters' reaction to Kibigori, a settlement just below the Nandi escarpment (Kenya).[13] To these porters this entrepot was also popularly known as Bagamoyo. Here they happily foresaw compensation for all the hardships they had endured (Fig. 3.1).

Evidence in form of song and testimony has been recorded to support the *wapagazi* theory of the origin of the name Bagamoyo. One experienced traveller observed: "Sometimes he can hear already from far the exciting shout "Bagamoyo,

---

[6]Hans Cory, 'Local Government Bagamoyo', *Hans Cory MS 122* and Hans Cory, 'Bagamoyo Zigua figurines', *Hans Cory MS 39A* (The University of Dar es Salaam); 'History before British occupation', *Bagamoyo District Book.*

[7]Charles Sacleux, *Dictionnaire Français-Swahili* (Zanzibar, Paris: Institut d'ethnologie, 1891), p. 85; Carl Velten, *Suaheli-Wörterbuch* (Berlin: Selbstverl Verf., 1910), vol. I, p. 33, translated 'B(w)agamoyo' as 'Herzberuhiger' [calming the heart]; Rev. Dr. Ludwig Krapf, *A Dictionary of the Swahili Language* (London: Trubner, 1882), pp. 19, 31, noted: "*bwaga* [. . .] to cast down what one has carried" and "*baga* [. . .] to carry for wages."

[8]Charles W. Rechenbach, *Swahili-English Dictionary* (Washington DC: Catholic University of America Press, 1967), p. 41.

[9]Frederick Johnson, *A Standard Swahili-English Dictionary* (Oxford: Oxford University Press, 1939), p. 43.

[10]Sacleux, *Dictionnaire*, p. 85. Presumably this story was told to Father Sacleux while he sojourned at Bagamoyo but subsequent research questions its veracity.

[11]' Bagamoyo', File 99.9 Ref. 59/2/5, Antiquities Library, National Museum, Dar es Salaam.

[12]'Histoire de Kabuera, jeune anthropophage du Manyema' (1885), in *Rapports divers du P. Guilleme* (1887–1889), C 19, p. 437, White Fathers Archives (Rome).

[13]C. H. Stigand, *The Land of Zinj* (London: Frank Cass, 1913), p. 289.

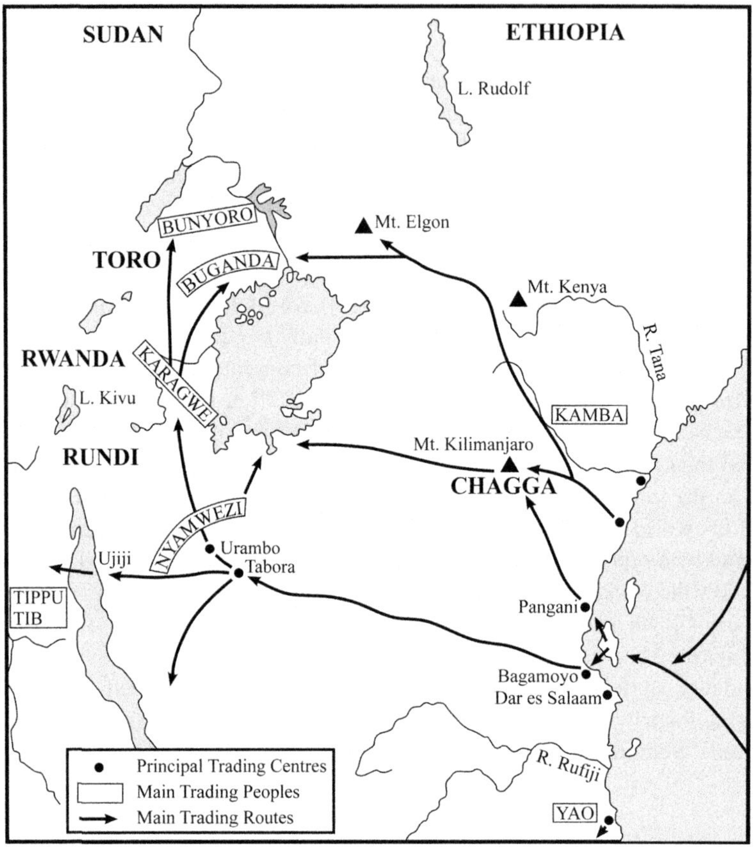

Fig. 3.1 Principle nineteenth century trade routes in East Africa, showing Bagamoyo, Zanzibar and Tabora mentioned by the author and also the location of the four other places called 'Bagamoyo' mentioned by him: Kibigory (A), Handeni town (1), east of Ndala (2), and south of Tanga (3) (map made by the Editor)

Bagamoyo."[14] It was a morale booster – a word of comfort, of encouragement and of the hope that they would soon reach the beautiful coast. As they journeyed eastwards, they often praised Bagamoyo – its promise and its sheer vitality[15]:

Be happy, my soul, let go all worries
soon the place of your yearnings is reached

[14]Leue, 'Bagamoyo', p. 11. See also: W. Griffith, Unyangwela to Urambo, 24 August 1789, Box 3, Journals: Central Africa, 1879, no. 9, London Missionary Society Archives, for another observation regarding songs about Bagamoyo, caravans, etc.

[15]Leue, 'Bagamoyo', pp. 11–12. This translation is also cited in Walter T. Brown, 'Bagamoyo: An Historical Introduction', *Tanzania Notes and Records*, 71 (1970), pp. 69–83. For the meaning of two words mentioned in this song: for 'dhow' see Fig. 3.3, and for 'ngoma' see later in this chapter.

Fig. 3.2 The slaves of Sirboko – an ivory merchant working for some Arabs at Zanzibar – carrying fuel and cutting rice for Speke's expedition in 'Unyamuezi', the inland area under the influence of Nyamwezi traders (Based on engraving in Speke's *Journal*, 1863 (John Hanning Speke, *Journal of the Discovery of the Source of the Nile* (Edinburgh, London: William Blackwell and sons, 1863), ch. 5, n.p.), redrawn by the Editor)

the town of palms – Bagamoyo.
Far away, how was my heart aching
when I was thinking of you, you pearl,
you place of happiness, Bagamoyo.

There the women wear their hair parted
You can drink palm wine all year round
In the garden of love, Bagamoyo.
The dhows arrive with streaming sails
and take aboard the treasures of Uleias
in the harbour of Bagamoyo.

Oh, what delight to see the *ngomas*
where the lovely girls are swaying in dance
at night in Bagamoyo.
Be quiet my heart, all worries are gone.
the drum beats and with rejoicing
we are reaching Bagamoyo.

Since members of various ethnic groups served as *wapagazi*, the discussion necessarily turns to the question of who was responsible for the origin of the name. Although only in the year 1839 do we first see a written reference to Bagamoyo,[16]

[16]E. Burgess, 'Letters from Mr. Burgess, dated 11 September 1839', *Missionary Herald of the American Board of Commissioners*, 87 (1840), p. 118. A thorough survey of Portuguese maps has revealed that the coastal area south of Utondwe was designated as Terra de San Raphael.

the earliest specific date is to be found in one of the town's graveyards: A. H. 1208 – A. D. 1793/94.[17] This date coincides with the period during which the long-distance caravan trade from the interior commenced. Thus, while it is generally asserted that Bagamoyo is the combination of two Swahili words *bwaga* and *moyo*, there is always the chance that the words have different linguistic origin. The hypothesis that the Nyamwezi were the first to arrive at the coast opens up the possibility that Nyamwezi might be the language of origin. But while the noun *moyo* also means 'heart' in Nyamwezi (and some other Bantu languages), the verb *bwaga* presents a problem. Edmund Dahl mentions *bagika* (−vaga, verb) as 'auseinanderstieben', i.e., 'wicked' or 'evil'; and *baga* (masculine, adjective) as 'schlecht, böse', i.e., 'spreading apart' – and thus sheds little light on the problem.[18] Much more detailed work will have to be undertaken by linguists and glottochronologists before a thorough exploration can be completed. Thus, while *bwaga-moyo* will be accepted as Swahili in origin, an alternate origin should not be rejected.

A further complication in the analysis of the origin of the word occurs in the proliferation of sites called Bagamoyo.[19] For, although we have been concentrating on the town which is the subject of this study, at least eleven other 'Bagamoyo'-s indiscriminately dot the Tanzanian mainland. The origin of these names thus poses both logistical and historical questions. Were the sites perhaps named in honour of Bagamoyo, that is, perhaps by a person or persons who had visited or lived at the famous coastal entrepot, or perhaps even imposed by coastal merchants who set up provisional commercial outposts in the interior? The answer (more likely the answers) is extremely difficult to ascertain. Only a very few pre-1885 maps of the interior are detailed and it is only in the last two decades of the nineteenth century that we find cartographical and historical references to these other 'Bagamoyo'-s.[20] In fact, oral traditions indicate that at least a few of these sites were established before the German period in East Africa.

This researcher visited two of the 11 'Bagamoyo'-s and data on a third was gathered by a fellow researcher. Within the town of Handeni (Handeni District), the area surrounding a small bridge which connects two main sections of the town is called Bagamoyo. Among the all too few inhabitants with whom I had a chance to confer the general, albeit vague, consensus was that the name Bagamoyo was Zigua in origin and the meaning had 'something to do with laying down one's

[17]This date appears on a tombstone in Makaburini Mwana Makuka. A. H. means the Islamic year-numbering system, commemorating the 'Hijra' (migration of Muhammad from Mecca to Medina in 622 A. D.).

[18]Edmund Dahl, *Nyamwesi-Wörterbuch* (Hamburg: L. Freidrichsen, 1915), pp. 5, 6, 184.

[19]The repetition of coastal names in the interior is an intriguing but ignored aspect of East African history. Bagamoyo probably occurs more than any other name, but I know from personal observation that there is a Saadani village in Hehe (Irgina District) and a Kaole village by the Uluguru Mountains (Morogoro District).

[20]See for example, Lieut, Chas, Stewart Smith, 'Explorations in Zanzibar Dominions', Supplementary Papers, *Royal Geographical Society*, II, part II, pp. 125–126.

heart."[21] The second site is a mile east of Ndala (Tabora District) where I had hoped to pursue the possibility of a Nyamwezi origin. The historical explanation for the name was interesting: the village was named by Swahili and/or Arab merchants from the coast who, during the early nineteenth century, traded in the area.[22] This explanation supports in part the nineteenth century Nyamwezi leader Mirambo's statement that even the word Nyamwezi is not Nyamwezi: "The Arabs of Zanzibar and the Banyans and Hindis of Bagamoyo gave the name to all the countries around about here."[23]

The third Bagamoyo checked is a small coastal village just south of Tanga (Tanga District). A Digo[24] was interviewed and recalled that the site had nothing whatsoever to do with the people of 'the' Bagamoyo. On the contrary, the village had received its name "a long ago." An *ngoma* (dance) competition was held there and for some unknown reason the site has ever since been called Bagamoyo. Additional information concerning the "Bagamoyo wa Waarabu zamani" was also offered. It had received its name because of its position at the beginning and end of the slave route and therefore was "a resting place (Fig. 3.1)."[25]

Ideally, each 'Bagamoyo' should be visited before any conclusion is reached. This of course takes a great deal of time and money and proved an impossible undertaking for this researcher. My conclusion then is based on information such as the above and on talks with numerous Tanzanians who took a deep interest in this phase of a larger project. In summation, I propose that an East African coastal site (lat. 6° 26′ S) was designated as 'Bagamoyo' by Swahili-speaking *wapagazi* during the late eighteenth or early nineteenth century. The significance of the name was gleaned from a commercial framework and indeed the history of Bagamoyo constitutes a significant chapter in the economic expansion of East Africa.[26] Bagamoyo emerged as a symbol not merely of a dynamic era in East African history, but more importantly, of man's ability to satisfy his physical, material and spiritual expectations (Fig. 3.3).

[21]Interview with P. Nkayemka at Korogwe in October 1968.

[22]Interview with Mzee Amani at Ndala in January 1969.

[23]E. J. Southon, 'The History of Country and Peoples of Unyamwezi', in a report attached to Southon to Whitehouse, 28 March 1880, Urambo, Box 3, Folder 1, Jacket C, London Missionary Society Archives.

[24]Ethnic and linguistic group that lives on the coastal strip between Mombasa and Tanga.

[25]Personal communication with Leif Landberg of Kigombe (Tanga District) in June 1968. For a note on the 'Bagamoyo-Ngoma' see *Usambara Post*, 31 December 1908, p. 1. The sentence in Kiswahili means "Bagamoyo was Arabised long ago" (the Editor is grateful to Garth A. Myers for the translation).

[26]See Walter T. Brown, 'The Politics of Business: Relations between Zanzibar and Bagamoyo in the Late Nineteenth Century', *African Historical Studies*, 4, 3 (1971), pp. 631–643.

Fig. 3.3 Present-day panorama of dhows in Bagamoyo's fishing port (Photo courtesy of Adam Jones, Ph.D./Global Photo Archive/Flickr)

Chapter 4
'The Trees Are Yours': Nature, Toponymy and Politics in the Interpretation of Cultural Landscapes in Lusaka and Zanzibar

Garth Myers

Abstract This chapter argues for taking African cultural practice seriously in developing urban environmental ideas about toponymy in Africa's cities, with a focus on Lusaka and Zanzibar. I dig into the contrast, or apparent contrast, between the two cities in urban toponymy related to trees and neighbourhood names, and test it against the realities of the actually lived and produced socio-environments of these two cities in the contemporary context. Conceptually, I build from a 'new' cultural geography framework developed by Denis Cosgrove and interpretive methods developed by James Duncan to extend the analysis of intersections between cultures and environments in urban place-names and cultural practices-in-place. If we are to fully read urban environments on the continent, and their toponyms, we need to have a more nuanced framework than one which poses power versus counter-power as a mere duality in urban politics. Everyday cultural conceptions of urban environments and place-names are crucial, often hidden dimensions of the 'produced ecosystems' in Africa, and they offer a valuable lens on why a diversified and nuanced form of urban political ecology is needed for approaching the continent's cities.

Keywords Cultural landscape • Lusaka • Zanzibar • Urban Africa • Neighbourhood names • Tree names • Political ecology

> It can be truly said of Africa, trees hide a multitude of sins.[1]

[1]Eric Dutton, *The Planting of Trees and Shrubs: with Special Reference to Northern Rhodesia* (Lusaka: Government Printers, 1937), p. 43.

G. Myers (✉)
Paul E. Raether Distinguished Professor of Urban International Studies,
Trinity College, Hartford, CT, USA
e-mail: Garth.Myers@trincoll.edu

L. Bigon (eds.), *Place Names in Africa*, DOI 10.1007/978-3-319-32485-2_4

> In Zanzibar trees hold a very particular place in the thoughts and acts of the people [...] Tree worship is universal in Zanzibar.[2]

4.1 Introduction

Eric Dutton was both the 'brain-child'[3] behind the design and implementation of Lusaka as the new capital of British Northern Rhodesia (today's Zambia) in the 1930s and the 10-year development plan for Zanzibar a decade later. He had a longstanding interest in gardening, landscape architecture, and arboriculture, amongst his many talents honed over a 35 year career in the service of the British Empire.[4] He fancied himself a keen observer of African cultural geography, and wrote four books laden with those observations.[5]

In the two quotations with which I open this chapter, Dutton provides a launching pad for exploring an apparently dramatic contrast in the urban environments of the two African cities which he most impacted. In Lusaka, where Dutton oversaw the creation of a supposed Garden City for Africa, he strategised that trees were a way of hiding a "multitude of sins", a moral imposition on a "land of flies and want."[6] Northern Rhodesia's people, white and black, apparently needed instruction in every aspect of tree cultivation, as much as they needed a cityscape that would remind them of the Empire's might, an "answer in bricks and mortar", in Dutton's view.[7] In Zanzibar, by contrast, he claimed that trees were so important that the people 'worshipped' them. His introduction to a guidebook to plants there which he commissioned – and from which this second quotation is taken – was meant for newly arriving officials; Zanzibaris already knew what they were doing with trees and shrubs, as far as Dutton saw it, because of their 'worship.'

In his examination – and his production – of cultural geographies on the continent, Dutton also paid careful attention to toponymy. As a principal mountaineer in what was nearly the second recorded ascent of Mount Kenya in the 1920s, he gave official names to many of the mountain's features as a part of his unsuccessful attempt on the summit.[8] In Lusaka, and especially in Zanzibar, he used names

[2]Eric Dutton, 'Introduction', in Robert O. Williams, *The Useful and Ornamental Plants of Zanzibar and Pemba* (Zanzibar: Government of Zanzibar, 1949), pp. 5–32 (p. 32).

[3]Elspeth Huxley, 'Introduction', in Eric Dutton, *The Night of the Hyena* (Unpublished memoirs, on microfilm, Oxford: Rhodes House Library).

[4]Garth Myers, *Verandahs of Power: Colonialism and Space in Urban Africa* (Syracuse: Syracuse University Press, 2003), pp. 18–23.

[5]Dutton, *The Planting of Trees* (and the unpublished memoirs, *Night of the Hyena*); *The Basuto of Basutoland* (London: Jonathan Cape, 1925); *Kenya Mountain* (London: Jonathan Cape, 1929) and *Lillibullero, or The Golden Road* (Zanzibar: Privately Published, 1944).

[6]Dutton, *Night of the Hyena*.

[7]Dutton, in a letter to Joseph Oldham, cited in Myers, *Verandahs of Power*, p. xi.

[8]Dutton, *Kenya Mountain*.

for places – buildings, streets, or neighbourhoods – to construct his vision of the political culture of British colonialism in Africa.[9] Oddly enough, though, Dutton's love of and interest in trees never met, at least in print, with his love of and interest in toponymy. If it had, he might have offered further evidence of the differences between the two places which these quotations at the head of the chapter suggest. Alas, his "unashamed colonialist"[10] mindset would have blindered the insights he might have been capable of. One sees this mindset, for instance, in the patronising tone of the phrase, "tree worship is universal in Zanzibar", when, of course, he would have known that in religious terms over 90 % of Zanzibar's population were Muslims, and even the rest – Christians, Hindus, and a few Sikhs – would have hardly 'worshipped' trees.

In this chapter, I want to dig into this contrast, or apparent contrast, hinted in the two Dutton quotations, and to test it against the realities of the actually lived and produced socio-environments of these two cities into the contemporary context. Conceptually, I take my cues from what was once the 'new' cultural geography, via a framework borrowed from a Denis Cosgrove essay. As Cosgrove put it in this nearly 30 year-old piece, "the many-layered meanings of symbolic landscapes await geographical decoding."[11] They still do. Working from the Gramscian ideas of Raymond Williams, Cosgrove suggested the possibilities of a decoding framework that would identify dominant, residual, emergent and excluded landscapes. Since Cosgrove's essay did not really aim at methodological rigour, I seek to do a slice of this decoding with methods suggested in James Duncan's *The City as Text*. Duncan's work was crucial in turning geographers toward the analysis of "landscape as a signifying system."[12] His analysis rested on hermeneutic research, in scholarly interpretation of "what a landscape signifies to those who produce, reproduce or transform it."[13] In doing so, Duncan examined the "tropes which allow a landscape to act as a sign system" through *allegory* (wherein "people, particularly powerful people, tell morally charged stories about themselves, the social relations within their community, and their relations to a divine order"), *synecdoche* ("the employment of a part to stand for the whole or the whole to stand for the part"), *metonymy* (where "a word or an icon stands for something else") or simple *repetition*.[14] In this chapter, I suggest interpretations of the allegorical, synecdochic, metonymic, and repetitive signs in Lusaka and Zanzibar's neighbourhood place-names and relationships with trees, in dominant, residual, emergent and excluded

---

[9]Garth Myers, 'Naming and Placing the Other: Power and the Urban Landscape in Zanzibar', *Tijdschrift voor Economische en Sociale Geographie*, 87, 3 (1996), pp. 237–246.

[10]Elspeth Huxley described him this way, in her introduction to Dutton, *Night of the Hyena*, p. ix.

[11]Denis Cosgrove, 'Geography is Everywhere: Culture and Symbolism in Human Landscapes', in Derek Gregory and Rex Walford (eds), *Horizons in Human Geography* (Basingstoke: Macmillan Education, 1989), pp. 118–135.

[12]James Duncan, *The City as Text* (Cambridge: Cambridge University Press, 1990), p. 17.

[13]Duncan, *City as Text*, p. 17.

[14]Duncan, *City as Text*, p. 20.

expressions. Since that is obviously a massive undertaking, I warn at the outset that the chapter is suggestive of a great many avenues for further deployment of these conceptual tools.

4.2 Nature, Toponymy and Politics in the Interpretation of African Cultural Cityscapes

Some of what we will see in the interpretation of the case studies is, in a sense, a clash between official and non-official discourses of representation via toponymy and trees, but with a broader and more nuanced sense of the spectrum of discourses across dominant, residual, emergent and excluded voices of the two towns. This study parallels Bekker and Therborn's study of capital cities in Africa, both in their analysis of how "the nation state projects its power through the urban landscape and spatial layout of the capital city" as well as its "architecture [. . .] public monuments and the names of its streets and public spaces", and in their exploration of what they call 'counter-power,' but, thinking of Cosgrove, in all of its 'multi-layered' character, rather than simply in an 'us-versus-them' form.[15]

I contend that if we take the old new cultural geography of Cosgrove and Duncan out for a more explicitly political drive, if you will, linking it with both political theories regarding states in urban landscapes and urban political ecology, we can deepen the significance of the contrast between Lusaka and Zanzibar which follows. In *Seeing Like a State*, James Scott examined how modern states sought to overcome the "spatial unintelligibility" of organically developed medieval cities in order to "make urban geography transparently legible."[16] Empires, like the British empire in Northern Rhodesia (and to a lesser extent Zanzibar), would clearly "find it symbolically useful to have [. . .] camps and towns laid out according to formula as a stamp of [the empires'] order and authority." Scott argued that little had changed from medieval and colonial times in the drive of states for this legibility in contemporary cities, at least as far as motivations go. What he called the state's "project of legibility" would, he contended, never be "fully realized", because of the "resistance of its subjects."[17] I argue that Cosgrove's refraction of alternatives to the dominant project of legibility, in residual, emergent, and excluded landscapes provides a more nuanced understanding of the terrain of public and popular responses to the dominant project. One can see these responses via the tools for analysing the rhetoric of landscape which Duncan provided – as I seek to show in both Lusaka and Zanzibar, a city with far less of the imprint of colonial power in its place-names or trees.

[15]Simon Bekker and Gören Therborn, 'Introduction', in Simon Bekker and Goren Therborn (eds), *Capital Cities in Africa: Power and Powerlessness* (Cape Town: HSRC Press, 2012), pp. 1–6 (p. 1).

[16]James Scott, *Seeing Like a State* (New Haven: Yale University Press, 1998), p. 55.

[17]Scott, *Seeing*, p. 77.

Furthermore, I want to draw attention to the mundane and the everyday in this legibility and its contestation. Like Scott, Benedict Anderson "draws attention to the need for regular cultural practices to produce and reproduce the significance of national identities"[18] in shaping what he famously termed the imagined communities of nationalism projects. But there are regular cultural practices which counter this agenda. Homi Bhabha noted that "counter-narratives of the nation that continually evoke and erase its totalizing boundaries – both actual and conceptual – disturb those ideological manoeuvres through which 'imagined communities' are given essentialist identities."[19] Moreover, many of these disturbances are in fact, mundane, everyday, and banal; rather than harbingers of resistance, cultural practices such as place-naming, or imbuing trees with cultural significance (potentially, but not exclusively, through place names) are often basic manifestations of the production of cultural landscapes.

If the thoughts above help to show how we might merge the old new cultural geography into political theory in this chapter, there is still the matter of nature, or, more specifically, urban natures. Let us use the example of urban trees. Jones and Cloke tell us that the "fate of trees is often emblematic of the wider environment",[20] including the political-economic environment, since "tree-landscapes are bound up with all manner of powerful cultural constructions, not least national identity."[21] Keil and Graham argue that "in the post-Fordist city, nature has become a major discursive element in the production of urban space",[22] and trees – perhaps especially avenue trees – are a key element of this. Urban political ecologists, in particular, have drawn scholarly attention to the co-production of nature and society, emphasising capitalist relations within this co-production. Urban political ecologists seek to "disrupt the idea of the city as the anti-thesis of nature", building on Marxist urban theory to see cities as socio-natural ecosystems produced in the power relationships of capitalism.[23] Urban political ecology's (UPE) "historical materialist roots are much more evident" than those of rural-focused political ecology in Africa, or of the even thinner base of work on urban cultural landscapes

---

[18]Joe Painter and Alex Jeffrey, *Political Geography: an Introduction to Space and Power*, second edition (London: Sage Publications, 2009), p. 154.

[19]Homi Bhabha, *The Location of Culture* (London: Routledge, 1994), p. 149.

[20]Owain Jones and Paul Cloke, *Tree Cultures: The Place of Trees and Trees in their Place* (London: Berg, 2002), p. 2.

[21]Jones and Cloke, *Tree Cultures*, p. 20.

[22]Roger Keil and Stephen Graham, 'Reasserting Nature: Constructing Urban Environments after Fordism', in Bruce Braun and Noel Castree (eds), *Remaking Reality: Nature at the Millennium* (London: Routledge, 1998), pp. 98–124 (p. 119).

[23]Alex Loftus, *Everyday Environmentalism: Creating an Urban Political Ecology* (Minneapolis: University of Minnesota Press, 2012), p. 3. See also: Maria Kaika and Erik Swyngedouw, 'The Urbanization of Nature: Great Promises, Impasse, and New Beginnings', in Gavin Bridge and Sophie Watson (eds), *The New Companion to the City* (Oxford: Wiley-Blackwell, 2011), pp. 96–107; Nik Heynen, Maria Kaika, and Erik Swyngedouw (eds), *In the Nature of Cities: Urban Political Ecology and the Politics of Urban Metabolism* (London: Routledge, 2006); Roger Keil, 'Urban Political Ecology: A Progress Report', *Urban Geography*, 26 (2005), pp. 723–738.

on the continent.[24] Because urban political ecology emerged largely outside of the continent, and particularly in reference to the West's cities, it is not surprising to find that UPE which builds specifically from African theoretical or conceptual bases, outside of South Africa, is thin on the ground, and African voices of UPE are nearly unheard.[25] Exceptions exist in modestly expanding number, but the small volume of UPE in urban Africa is made particularly striking by the evident potential for scholarly work examining urban environments from a variety of perspectives and building from African conceptualisations, given the magnitude of urban environmental-political challenges and conflicts.[26] UPE has thus far failed to grapple with the environmental philosophies and socio-natural constructions of African urban communities. Examination of African urban perspectives on trees and toponymy, as a means for fostering further understanding the rhetorical flows in the urban landscape, can thus produce a valuable contribution to the pluralisation of UPE.

4.3 Neighbourhood Names and Natures in Lusaka and Zanzibar

Before we can unpack the environmental philosophies or socio-natural perspectives of ordinary urbanites, though, we need a thorough empirical base in the toponymy. I have studied Zanzibari place names for 25 years, and place-names of Lusaka for 15 years. Although there are fascinating characteristics to the naming of monuments or streets, the strongest and most enlightening base for rigorous comparison – particularly given the combined interests of this chapter in both place-names and arboriculture – lies with neighbourhood names. I developed a sample of the official and commonly deployed names for neighbourhoods in the two cities. Both lists are as comprehensive as possible, 122 names from Zanzibar and 63 from Lusaka. Despite its much larger population size (1.7 million as opposed to 450,000), Lusaka has fewer names in the sample, largely because most of its neighbourhoods, as they expand to form new sections, do not often take on new names but rather add on English-language descriptors (such as 'Woodlands South Extension' or 'Kamanga

[24]Loftus, *Everyday*, p. 5; Mary Lawhon, Henrik Ernstson, and Jonathan Silver, 'Provincializing Urban Political Ecology: Towards a Situated UPE through African Urbanism', *Antipode* (2014) (forthcoming).

[25]Garth Myers, 'Political Ecology and Urbanization: Zanzibar's Construction Materials Industry', *Journal of Modern African Studies*, 37, 1 (1999), pp. 83–108; Garth Myers, *Disposable Cities: Garbage, Governance, and Sustainable Development in Urban Africa* (Aldershot, UK: Ashgate, 2005); Jeremia Njeru, 'The Urban Political Ecology of Plastic Bag Waste Problems in Nairobi, Kenya', *Geoforum*, 37 (2006), pp. 1046–1058.

[26]Tania Katzschner, 'Cape Flats Nature: Rethinking Urban Ecologies', in Lesley Green (ed.), *Contested Ecologies: Dialogues in the South on Nature and Knowledge* (Cape Town: HSRC Press, 2013), pp. 202–226.

Table 4.1 Neighbourhood names of Zanzibar and Lusaka compared

Category of names	Zanzibar (pct)	Lusaka (pct)
European (English) names	0 (0)	7 (42.9)
White settlers' names	0 (0)	7 (11.1)
Tree-related English names	0 (0)	5 (7.9)
Indigenous tree-related names	30 (24.6)	0 (0)
Cultural-historical names, narratives	34 (27.9)	17 (27)
Everyday economic activity	14 (11.5)	0 (0)
Land ownership	5 (4.2)	0 (0)[a]
Natural features (other than trees)	26 (21.7)	6 (10)
Political names	13 (10.7)	2 (3.2)
Other/undetermined	0 (0)	11 (17.5)

[a]None of the Lusaka neighborhood place names are derived from land ownership *other* than those that are derived from the personal names of white settler landowners, a separate category above

Overspill Area Five'), and I limited the sample names in these to the original neighbourhoods (e.g., 'Woodlands' or 'Kamanga').

As Table 4.1 makes plain, there are a few similarities and yet striking differences in the comparative toponym-geographies of neighbourhoods in Lusaka and Zanzibar. Among the differences, three stand out. First, a plurality (nearly 45 % per cent) of Lusaka's neighbourhoods are English names, and a little more than one-fourth of these are place-names derived from the names of the European settlers on those lands during the colonial era; by contrast, not one neighbourhood of Zanzibar is in English, and – perhaps obviously, given the lack of European settlers there – no neighbourhoods are named for Europeans.

Second, despite both cities having experienced decades of socialist rule after the end of British colonialism, only in Zanzibar do we see neighbourhoods which were named or renamed in a bout of revolutionary fervour in the 1960s. Many honour allied states of the Zanzibar Revolution of 1964 or allies in the fight against apartheid South Africa: Angola, Russia (*Urusi*), Mozambique (*Msumbiji*), *Muungano* (The Union (of Tanganyika with Zanzibar to form Tanzania)), and Nyerere (first President of that Union). By contrast, while Lusaka has 'Kaunda Square' (named for independent Zambia's first President), the only other neighbourhood with an explicitly political name is a new overspill area from Chelston that was, in 2009, designated with the name 'Obama.'

Third, and most pertinently, nearly 25 % of all neighbourhoods in Zanzibar have a tree in the name, while the only Lusaka neighbourhoods with trees in them (five, or less than 8 %) are all in English (as in Thorn Park, Woodlands, or Meanwood). Lusaka's avenue trees came from all over Africa and around the world. Clearly, 100 years after its creation and nearly 80 years after its re-launch as a garden city, Lusaka remains something of an alien imposition. Its treescapes certainly are, given the lack of indigenous names, commemorating trees (there is one which names a flower from a tree, but not the tree). Neighbourhoods which date to the colonial era that bear the names of trees are entirely within what were the European areas of the city. The three newly developing suburban communities that go by the name of 'Meanwood' (Chamba (Marijuana) Valley, Vorna Valley, and Ibex Hill) are on land which belonged to the white settler, Abraham Galoun, whose descendants long ran 'Galounia Farms' (the sellers of the Meanwood properties). The only indigenous, peri-urban informal settlement to have a name related to this colonial-origin tree-culture project is 'Garden', and this is a bit of a playful cultural joke: Garden is a neighbourhood built illegally in the sewage outfall of the city treatment facility.

In Zanzibar, the plentiful array of trees represented in the city's indigenous names for neighbourhoods surely betray a much greater significance for trees in the local urban culture as it developed. Mango, Baobab, Indian Almond, Jackfruit, Banana, Kapok, Wild Kapok, Neem, Rambutan, Saman, Tamarind, Mandarin Orange – tree after tree is a part or all of one-quarter of all neighbourhood names.

How should we interpret these three differences? They immediately suggest significant contrasts in the produced cultural landscapes of the two cities. The first difference might be termed the rootedness or indigeneity issue. Aside from the small populations of Lenje or Soli peoples present in the Lusaka area in 1913 when the township of Lusaka was officially established, the vast majority of Lusaka's population came to the city from other parts of Zambia or from abroad. Although Chinyanja serves as the lingua franca for many of its residents, English is increasingly a second such common language among Lusaka people. There are a few places with Soli or Lenje names (like 'Lusaka' itself – Lusaaka was a local Lenje leader in 1913; and 'Chilenje', i.e., place of the Lenje, is one of its oldest African areas; alongside Soli place-names like 'Chainda', i.e., it has passed, or 'Kalingalinga', i.e., headless snake). But most of the city sits on the landscape like an outsider, toponymically. The new areas labelled as 'overspill area number 1, 2, 3', and so forth, or 'extension', or 'south extension' add to this sense.

By contrast, Zanzibar has been and remains overwhelmingly an indigenous African-Swahili city; it predates the British colonial period by 200 years as a city, and six centuries as a settlement of some sort. The British imposition toponymically is virtually non-existent. Only Dutton's creation of the name 'Raha Leo' has stood the test of time as a European-imposed neighbourhood name. Ironically, the tiny neighbourhood known as 'Mitiulaya' (European Trees) designates a stately allee along both sides of the small street at the neighbourhood's centre; although the trees were planted during the British colonial period, the chief officers in charge of tree planting, even in the 1930s, were Goanese, and the Saman or Summer Rain Trees are spectacular and massive South American shade trees. There are a few references in neighbourhood names to Omani or Indian landowners – 'Kulateni' (Qulatein family), 'Kijambia' (the Omani ceremonial knife), 'Kwahani' (Land of Khan), or 'Kwaalinatoo' (Land of Ali Nathoo) – but the long dominance of Swahili working-class and peasant communities in the city and its suburbs is very evident in the place-names, in stark contrast to the colonial legacy evident in so many Lusaka neighbourhoods.

The political naming distinction is largely explicable by recourse to analysis of the socialist character of the two post-colonial states. While Zambia under

Kaunda was nominally socialist and operated under the banner of his 'Humanist' ideology as a single-party state from 1973 to 1990, Kaunda's style was highly pragmatic. The World Bank, after all, built the homes of Kaunda Square. Revolutionary Zanzibar, particularly under its first President, Abeid Amani Karume (1964–1972), was a staunchly socialist police state when neighbourhoods like 'Urusi', 'Muungano', 'Msumbiji' or 'Nyerere' were built, down to the menacing operations of the ruling Afro-Shirazi Party in those and other neighbourhoods. The model homes of upper Kikwajuni built immediately after the Revolution are called 'Kikwajuni GDR' because they were built by engineers from communist East Germany.

The most intriguing difference lies with the tree names. In creating Lusaka's Garden City, Dutton brought trees and shrubs from around the world into the *miombo* and *munga* woodlands of central Zambia. Some 60 species of trees and shrubs were imported from Australia, with another 35 coming from northern Tanganyika, eight from Kenya, and three from the Belgian Congo. Still others came from India, Brazil, or Southeast Asia. African trees utilised included a few species native to Northern Rhodesia, too, such as the Black Wattle, African Tulip Tree, Khaya (African Mahogany), or Flame Tree.[27] But Toon trees, Jacarandas, and various species of cassia and fig became by far the most prominent trees of the capital area's public face, in avenue trees.

Dutton also sought to expand the tree culture of Britain more broadly in the colony of Northern Rhodesia, in white settlers' private gardens and yards, "in the belief that the planting of ornamental trees and shrubs is of great importance to young townships, as well as to individuals, and should be encouraged in every way."[28] This use of trees was fundamentally bound up with the importance of private landed property, albeit at a smaller scale than the lordly estates of Yorkshire that Dutton grew up around. In discussing the creation of open home gardens without hedges or fences, as an apparent trend of the time in southern Africa, Dutton sought to emphatically undercut it: "It is a matter of opinion, but my own vote goes to a hedge, as high and as impenetrable as possible" to mark the property and the boundary of the garden.[29]

In the 1950s white residents in Lusaka built whites-only autonomous townships around it (Kabulonga, Roma, and Olympia); they planted many of these same exotic ornamental species Dutton had recommended on their plots, and then private white developers of those areas planted them on the streets of these townships. Now that these areas are dominated by a black Zambian middle-class or elite, it remains the case that avenues in them are lined by exotic jacaranda, palms, and the like, and many similarly foreign ornamentals are planted inside the cement wall fortresses of their plots. But in the 'unauthorised areas' locally known as 'compounds', the informal settlements where two-thirds of today's Lusaka population can be found,

---

[27] Dutton, *Planting of Trees*, pp. 130–136.

[28] Dutton, *Planting of Trees*, p. iii.

[29] Dutton, *Planting of Trees*, p. iii.

there is little tree cover other than small fruit trees and shrubs; most plants in these areas are crop plants.[30] The *miombo* and *munga* treescape has been largely cleared away from the urban context. Hence it is not really plausible to have had African-origin trees in the names of peri-urban compounds since the areas were often treeless to start with – this despite the general sense visitors may often have of Lusaka as a very 'green' city with a lot of trees.

Zanzibar is in a (dry) tropical forest biome, in contrast with the wooded savanna ecosystem of the Lusaka rea prior to urbanisation. It makes sense that it has a fuller assortment of trees. When we look inside those Zanzibari neighbourhood tree names a little, though, what we see is not worship at all, but the everyday life of working class and peasant urban communities. First of all, nearly every tree named in a Zanzibar neighbourhood is also an exotic species, one brought either by the British or by other European, Indian, or Arab traders over the centuries. Many of these – and varieties of palms or mangoes, most notably – have grown in such abundance, with many species and sub-species among these families present in the isles for so many years, though, that most urbanites would have found these trees in situ upon their arrival to the town even in the eighteenth or nineteenth century BCE.

Ten Zanzibari neighbourhood names have mango trees in them. Mango trees grow tall and broad, providing extensive shade; they thus become focal points for cultural and economic activity. Mango trees (*Mwembe*, plural *Miembe*) that had shade areas for sail-mending (*Mwembe-tanga*), advice-giving (*Mwembe-shauri*), coir-rope-making (*Mwembe-makumbi*), small-scale sales of groundnuts (*Mwembe-njugu*), fish (*Kiembe-samaki*), or sweets (*Mwembe-ladu*), and for pregnant women to rest by the maternity hospital (*Mwembe-mimba*). Most other trees named in neighbourhood toponymy are likewise large shade trees (kapok, baobab, jack-fruit, Indian almond), or even medium-height trees known for throwing shade (*Neem/Mwarubaini*), where people would gather. This is exactly what urban political ecologists mean by 'socio-nature': the trees and the culture are co-produced, and co-reproduced. The same 'socio-natural co-production' of the landscape happens in Lusaka, of course, but it requires more indirect interpretation to observe in the toponyms. But this culture-nature interface also points us toward the key commonalty between the two cases, and something very much absent from the UPE literature in reference to Africa: the prominence of indigenous cultural-historical narratives in neighbourhood naming.

Both Zanzibar and Lusaka have produced many neighbourhood names out of African cultural processes, events, rituals or stories. In some cases, the origin of these has been lost to residents in them, but others are marks of culture and local society embedded in the character of the community, or clear statements

[30]Dirk Jaeger and J. Dennis Hackabay, 'The Garden City of Lusaka: Urban Agriculture', in Geoffrey Williams (ed.), *Lusaka and its Environs: A Geographical Study of a Planned Capital City in Tropical Africa* (Lusaka: Zambia Geographical Association, 1986), pp. 267–277; Garth Myers, 'The Unauthorized City: Late Colonial Lusaka and Post-Colonial Geography', *Singapore Journal of Tropical Geography*, 27, 3 (2006), pp. 289–308.

of profound meaning politically. In Lusaka, one of the earliest illegal settlements took the name 'Chawama' (It is Beautiful), as a statement of defiance. A more recent compound makes a more pointed, and stark, political claim through its name: 'Chibolya' (Where Useless Stuff is). But a great many names are playful, funny, touching, or basic portraits of culture in action: 'Chilanga' (Staring), 'Kabwata' (People Who Talk Too Much), 'Kalikiliki' (Boxing/Fighting Place), 'Kamanga' (named for a local bricklayer who was the first illegal occupier of the settlement), 'Libala' (Marriage), 'Munali' (You Haven't Finished Talking), or 'Mutendere' (Peace). Even the compounds named for their white settler owners have been twisted in playful ways. G.B. Marapodi, an Italian cement maker in the city's early days, is remembered in the compound 'Marapodi' which bears his name; and also in the neighbouring compound of 'Mandevu' (Beards): a religious Jew, Marapodi maintained a substantial beard. His daughter Elizabeth also shows up, in Villa Elizabeta, and his wife is memorialised in the compound of 'The Mrs.', Zambianised as 'Misisi.'

This same memorialisation of the everyday shines through in many Zanzibari names. We have already seen this with the many tree-based names where a tree is paired with the activity under it. As in Lusaka, there are sharp commentaries enmeshed in some names, as in 'Daraja Bovu' (Broken Bridge, adjacent to a frequently flooded bridge structure), 'Uholanzi' (Holland, named because of frequent flooding), or 'Sogea' (Pack It In, a dense informal settlement). But there is also a Piece of a Road Ballast ('Baraste Kipande'), Lucky Coconut Pudding ('Bumbwisudi'), a Place to be Carried ('Chukwani'), a Place for Moonshine Liquor ('Gongoni'), a place to play Hide and Go Seek ('Kajificheni') and a place You Have to Love ('Mpendae'). Places of work and community activity are frequently named for the work or activity: 'Sokomuhogo' (Cassava Market), 'Gulioni' (At the Market), 'Karakana' (Workshop), 'Kama' (Milking), 'Mazizini' (At the Cattle Pens), 'Fuoni' (At the Clothes-washing Place), or 'Kinuni' (At the Grinding Mill).

Thus the place-names of both cities often enact the everyday socio-nature of the people, and in ways which are suggestive of local perspectives on nature, politics and cultural practice. What remains to be articulated are the threads back to the conceptual framework I laid out earlier in the chapter. As I noted early on, I can really only hint at these, given the brevity expected of the chapter.

4.4 Analysis

It is hard to miss the allegory and synecdoche in Zanzibar and Lusaka's treescapes or toponymies, from its dominant, residual, emergent, and excluded places. To be sure, in both symbolic and material ways, the contemporary city of Lusaka reproduces the colonial one in biogeography and place names; perhaps the best term for it is to think of Lusaka as a neo-colonial city. The symbolism of this is perhaps nowhere more synecdochically put (e.g. where a part stands for a whole) than in the office of the (white) Vice President of Zambia, Guy Scott; in 2013, Scott was occupying the Lusaka City Council Nursery offices in Woodlands, just beyond the Golf Club, at the old colonial city boundary – the nursery where some of the seedlings of the avenue trees for the colonial Garden City, literally, were grown.

The stories the powerful people want to tell with trees and names have remained rather similar for many years in Lusaka, figuratively exoticising the cityscape, while ordering Africa into a European box of beautification. In the emergent form of this story, the part that most graphically stands for the whole is the elite suburban plot: what we are seeing is the privatisation of the cityscape, where each plot contains within its ever higher walls an allegorical forest of wealth. Most new gated communities take on English tree names, or hint at them, such as 'Silverest Gardens', playing off of the Latin 'silva' (tree). In Zanzibar, the dominant bloc's narratives on the landscape have shifted from the colonial era to a revolutionary socialist one, and slowly now to a more nuanced form – the best example of this is not a neighbourhood, but the main city hospital: built under Dutton's post-war development plan with an Indian elite's name, it became 'V.I. Lenin Hospital' after the 1964 Revolution, but was renamed again in the late 1990s as 'Mnazimmoja [One Coconut Palm Tree] Hospital.'

The residual elements are there in Lusaka in the stubborn manner in which the failure of the colonial garden city and the gross injustices of it stab the eyes of visitors and residents alike, as in 'Garden', with its gloriously ironic place-name, a poor, largely treeless informal settlement often flooded with sewage. In Zanzibar, the residual presence of the Omani Sultanate is implicit in some names, and in the plantation of so many large fruit trees all over what became the city – these were, in the main, Omani-owned plantations in the nineteenth century. Despite the revolutionaries' famous slogan, 'The trees are yours, the land is ours', though, in reality Zanzibaris made the Omani- or Indian-imported trees *theirs* most evidently in the names of their neighbourhoods.

The excluded stories are often quite literally hidden. One has to work to even enter Lusaka's 'Misisi' compound other than on foot, since it is essentially off of even an informal road grid despite being within a kilometre of downtown; 'Chibolya', the 'place where useless stuff is', is similarly hard to access. In Zanzibar, the whole stretch of neighbourhoods just across the formal boundary of Zanzibar City in suburban West District are colloquially known as the 'Gaza Strip', with subdivisions like 'Baghdad', where hundreds of residents had their homes destroyed by the Tanzanian army in an effort to impose a buffer zone around their base; but no official map will ever tell you where this 'Gaza Strip' is, or 'Baghdad.'

Dutton may not have had any notion of the contrast his written works provided between the tree cultures of these two cities. In any case he spoke with the voice of colonial power, a voice that was contested and overthrown in both cities, with more lasting success in Zanzibar, perhaps. Yet when we return to Duncan's idea of landscape as a signifying system, with all of its allegory, metonymy, synecdoche, and repetition, with Cosgrove's variations in mind, we see that much more than just a rhetoric of power versus counter-power is at play. State projects of legibility clearly involve toponymy, and, at least in Lusaka, the project also involved tree-planting.

But residual, emergent and excluded 're-writing' of these projects are quite evident. There is a need to attend to the mundane and the banal, rather than seeing all naming as part of a grand, essentialist contestation. In parallel fashion, my chapter aims at pluralising urban political ecology, by opening it up to exploration of the genuinely 'everyday' environmentalism of a broad array of residents as expressed in naming practices.

Chapter 5
Transnational Aspects in the History of Lagos: Place Names and Built Forms

Liora Bigon and Robert Home

Abstract It may seem curious that Lagos, a city of many millions, recognised as a world metropolis, should still be known by most of its Nigerian inhabitants among themselves as *Eko*. This chapter aims to trace the toponymic history of Lagos over the 400 years since its original settlement, making analytical connections between place naming and built form. From Oko to Eko, from Curamo to Onim, and from Onim to Lagos – each of these name transfers represents a layer in the rich and cosmopolitan past of the city. Through multiple primary and secondary sources (written and oral histories, cartography and architectural evidence), the shifting meaning of toponyms within the urban complex can be scrutinised. Transnational and multilateral aspects of sub-Saharan Africa's history will be highlighted, with the aim of exposing the complexities of the simultaneous usages in Lagos's names. These constitute a fertile ground for juxtaposed memories on the part of the involved agencies, ethnic groups, and political, economic and cultural powers, memories which sometimes challenge one another, and are sometimes complementary.

Keywords Lagos' toponymy • Nigeria's history • European cartography • Built forms • Urban planning • Yoruba • Cosmopolitan heritage

5.1 Introduction

Lagos metropolitan area – population estimated at more than ten million and projected to reach nineteen million by 2025 – is not only the largest city in Nigeria and in sub-Saharan Africa, but also one of the largest in the world, a

L. Bigon (✉)
The Institute of Western Cultures, The Hebrew University of Jerusalem, Jerusalem, Israel

Holon Institute of Technology, Holon, Israel
e-mail: liorin@hotmail.com

R. Home
Land Management, Anglia Ruskin University, Cambridge, UK
e-mail: Robert.Home@anglia.ac.uk

L. Bigon (eds.), *Place Names in Africa*, DOI 10.1007/978-3-319-32485-2_5

megacity.[1] Two competing views have characterised current critical research on its urban morphology and management. The Harvard Design School Project on the City, chaired by the Dutch architect-urbanist Rem Koolhaas,[2] sees the chaotic aspects of Lagos's present development as a series of self-regularising socio-economic structures beyond any formal surveillance. This approach sees no threat in urban chaos, rather highlights the new sense of cityness and its innovative and improvisational abilities, which have virtually escaped the Western discourse on modernity and conform to the challenges of the twentieth century. By contrast the 'developmentalist' approach in urban geography sees Lagos as symbolising anarchy, a kind of inconceivable hell that operates without any hope for improvement. Officially proclaimed by the UN in 1991 as the dirtiest city in the world, it is characterised by massive population growth, widespread poverty and slums, traffic jams, violent crime and poor access to safe drinking water and other services, constituting multiple vulnerabilities and risks for the inhabitants, and part of wider political and socio-economic problems in Nigeria.[3]

The idea of transferring the capital (first suggested by the first Governor-General of colonial Nigeria, Lugard[4]) closer to the geographical centre of Nigeria resulted in Abuja, a planned city in a federal capital territory, being declared Nigeria's capital in 1991. Lagos, however, remains one of Africa's most vibrant and cosmopolitan centres. A stranger might be surprised that the name *Lagos* is in everyday use by only a few, with the name *Eko* preferred among Lagos's urban majority, the Yoruba people.[5] This is with particular reference to Lagos Island, the original settlement and the present-day centre of the city, while the contemporary urban agglomeration is polycentric and expands over the adjacent mainland, but the island still serves as the main seaport, business and administrative district, and a prestigious living area.

An attempt to understand this toponymic dualism takes us to the genealogy of Lagos's toponymy, its meaning and connotations, reflecting the physical and social development of the place. Connections between changing the settlement's name and its changing socio-political character will be highlighted, stressing urban form

---

[1]Short parts were borrowed, with permission, from a previous publication: Liora Bigon, 'The Former Names of Lagos (Nigeria) in Historical Perspective', *Names: A Journal of Onomastics*, 59, 4 (2011), pp. 229–240 (www.maneyonline.com/nam). About Lagos' population see: UN-Habitat, *The State of African Cities, 2014:Re-Imagining Sustainable UrbanTransitions* (Nairobi: UN-Habitat, 2014), p. 103.

[2]Rem Koolhaas (et al., eds), *Mutations* (Bordeaux: arc en rêve, centre d'architecture, 2000).

[3]Matthew Gandy, 'Planning, Anti-Planning and the Infrastructure Crisis Facing Metropolitan Lagos', *Urban Studies*, 43, 2 (2006), pp. 371–396; Laurent Fourchard, 'Lagos, Koolhaas and Partisan Politics in Nigeria', *International Journal of Urban and Regional Research*, 35, 1 (2011), pp. 40–56.

[4]For Lugard's plans see: Rhodes House (Oxford), Lugard, MSS. Brit. Emp. S. 99, 1: 1901–1916, Revenues and expenditure, ff 238–49, 1913.

[5]Numbering about 50 million, the Yoruba are one of the largest ethnic groups in West Africa. Concentrated in the southwestern part of Nigeria, they make up more than 20 % of this country's total population, and are divided into numerous sub-ethnic groups – such as the Oyo, Ife, Egba, Awori and Ijebu – each with its own Yoruba dialect, towns and villages.

in terms of planning and architectural morphogenesis. In addition to toponyms *of* the urban complex, toponyms *within* the urban complex will be touched upon, such as names of quarters or streets, particularly if the latter names illuminate our main argument. Between the name *Eko*, one of the original names given to the initial settlement, and the current name *Lagos*, its European origin has been naturalised in local use and official cartography over about 400 years. During most of this period the two names were used simultaneously, as well as several other names such as *Oko*, *Curamo* and *Onim*, indicating rich and competing narratives in the dynamic spatial configuration of the settlement. Various political, economic, ethnic and cultural forces left their landmark on the place-names given to the settlement and its landscapes and built form.

Our approach goes beyond what is called 'area studies', which tends to treat Africa as a unique 'out there' from the perspective of the global North. Rather than place Africa at the centre or Europe at the periphery, we identify the transnational aspects that might be called 'connected histories' or 'crossed histoy'[6] (see Introduction). The crossing of the borders of naming and spatialising processes in Lagos does not include only local site-related intra- or inter-regional influences in West Africa, but also multilateral, transatlantic influences across three continents: Africa, Europe and South America. 'Transnational' here refers to more than an informal crossing of formal borders of modern national states (which did not exist over most of the period in question), but to the many toponymic and physical influences arising from the encounter of multiple ethno-political and economic entities. To borrow the words of the editors of *The Palgrave Dictionary of Transnational History*: "We are interested in links and flows, and want to track people, ideas, products, processes and patterns that operate over, across, through, beyond, above, under, or in-between polities and societies."[7]

An approach put forward by Catherine Coquery-Vidrovitch will be embraced. Recognising the historical depth of African cities, Coquery-Vidrovitch argues the inadequacy of a typology classifying them according to a dominant feature at a given historical point. She instead suggests a chronological approach to African urbanisation which accepts a variety of developmental models through the relevant eras and long intermediary phases of interpenetration between them.[8] The relevance of the chronological approach is supported by the fact that each place name constitutes another layer in the conception of the place, allowing both names, *Eko* and *Lagos*, to persist to this day, revealing corresponding, complementary or

[6]See for more: Sanjay Subrahmanyam, 'Connected Histories: Notes towards a Reconfiguration of Early Modern Eurasia', *Modern Asian Studies*, 31, 3 (1997), pp. 735–762; Michael Werner and Bénédicte Zimmermann, 'Beyond Comparison: *histoire croisée* and the Challenge of Reflexivity', *History and Theory*, 45, 1 (2006), pp. 30–50.

[7]Akira Iriye and Pierre-Yves Saunier, 'Introduction', in Akira Iriye and Pierre-Yves Saunier (eds), *The Palgrave Dictionary of Transnational History from the Mid-19th Century to the Present Day* (Basingstoke: Macmillan, 2009), pp. xvii-xx (p. xviii).

[8]Catherine Coquery-Vidrovitch, *The History of African Cities South of the Sahara: From the Origins to Colonization*, trans. Mary Baker (Princeton: Markus Wiener, 2005), 26.

even competing narratives and memories. This understanding accords with trends in toponymic research over the last twenty years. In attempting new directions in critical toponymic studies, Rose-Redwood, Alderman and Azaryahu argue that reliance on maps and gazetteers needs to be supplemented by an analysis of archival records, work in situ, interviews and other almost ethnographic methods.[9] This qualitative and mixed method also helps understand Lagos's history.

The series of successive names assigned to the settlement and its emerging quarters begins before an exact dating, a few hundred years ago, with the preliminary name *Oko*. The current name *Lagos* can be dated to the mid-nineteenth century with the establishment of British colonial rule. For much of the examined period the relevant city-texts were not systematically documented, requiring the use of multiple sources such as: fieldwork including an acquaintance with the site, interviews and collection of visual evidence; secondary and primary archival sources particularly from Nigeria and Britain; oral traditions; written/cartographic documentation of European visitors; and a century-old postcard photos. Limits of course exist in trying to reconstruct this toponymy of the hidden city-text, with almost nothing known on the beginning of the settlement because of perishable building materials in the tropical climate, termite destruction, and the impossibility of opening archaeological sites in the congested and multi-layered zones of the island, subject to post-colonial violent slum clearance schemes and urban regeneration.

In historiography the subject of Lagos's names is usually treated in passing, never by itself, within a broader literature predominantly of monographs on Lagos's history, its socio-political and urban history, and its architectural and town planning history.[10] Pierre Verger's 1959 paper deals exclusively with one of Lagos's former names,[11] although it follows the taxonomic-encyclopaedist research approach of the time of writing. Robin Law's 1983 work attempts to understand the West African 'slave coast' trade and its geo-political implications, with an excellent

[9]Reuben Rose-Redwood, Derek Alderman and Maoz Azaryahu, 'Geographies of Toponymic Inscription: New Directions in Critical Place-Name Studies', *Progress in Human Geography*, 34, 4 (2010), pp. 453–470 (pp. 456, 466).

[10]Respectively: A. B. Aderibigbe, 'Early History of Lagos to about 1850', in A. B. Aderibigbe (ed.), *Lagos: The Development of an African City* (London: Longman, 1975), pp. 1–25; Kristin Mann, *Slavery and the Birth of an African City: Lagos, 1760–1900* (Bloomington: Indiana University Press, 2007); Pauline Baker, *Urbanization and Political Change: The Politics of Lagos, 1917–1967* (Berkeley: University of California Press, 1974); Margaret Peil, *Lagos: The City is the People* (London: Belhaven Press, 1991); Eva Krapf-Askari, *Yoruba Towns and Cities* (Oxford: Oxford University Press, 1969); Kunle Akinsemoyin and Alan Vaughan-Richards, *Building Lagos* (Lagos: F. & A. Services, 1976); Liora Bigon, *A History of Urban Planning in Two West African Colonial Capitals: Residential Segregation in British Lagos and French Dakar (1850–1930)* (Lewiston: The Edwin Mellen Press, 2009).

[11]Pierre Verger, 'Notes on some Documents in which Lagos is Referred to by the Name 'Onim' and which Mention Relations between Onim and Brazil', *Nigeria Historical Society Journal*, 1 (1959), pp. 343–350.

though seemingly inadvertent regional toponymic analysis.[12] Nkiru Nzegwu's 1996 article is a lively discussion of Lagos's spatiality through certain Yoruba key notions about the urban sphere.[13] Lagos's toponyms are also discussed in passing by accounts of Yoruba missionary-educated intellectual elites in the late nineteenth and early twentieth century, attempting to document their mythological or historical past.[14] Different names are used in primary sources written by European traders, missionaries and diplomats, including travel diaries and cartographic data between 1500 and about 1890. We used, for example, a map of the coast of West Africa made by the Dutch cartographer Willem Bleau in 1635, or a Lagos map from 1885 drawn in remarkable scale and detail by the African surveyor W. T. C. Lawson. The memories of the Italian Consul in Lagos (Giambattista Scala) after 1862 were also useful, as was the series of postcard photographs taken by the French photographer Edmond Fortier, and the authors' photos. The contribution of this study is in bringing these primary and secondary sources from the background to the foreground of the analysis. Conversations between toponyms and urban forms are rare in historical studies of the global South, but inform most of the contributions in this pioneering volume.

5.2 From 'Oko' to 'Eko': The Initial Beginning

"The first man that built Isheri and settled there, in or about the year 1699, was a hunter, named Ogunfunminire, meaning 'the god of iron has given me success'. He was of the royal family of Ile-Ifẹ." This is what we are told by John B. Losi, an indigenous headmaster of one of the Roman Catholic schools in Lagos, and an early historian who collected and reassessed Yoruba oral traditions.[15] In Isheri, about twelve miles north of Lagos Island, the Awori, a Yoruba sub-group, established their first of several settlements, trying to escape contemporary political instability in the interior of Yorubaland. Losi's dating is unreliable, for any date before the fifth Oba of Lagos Ologun Kutere (on the throne in the 1780s and 1790s, as documented by European traders) is arbitrary.[16] The subsequent developments, however, are

[12]Robin Law, 'Trade and Politics behind the Slave Coast: The Lagoon Traffic and the Rise of Lagos, 1500–1800', *Journal of African History,* 24 (1983), pp. 321–348.

[13]Nkiru Nzegwu, 'Bypassing New York in Re-presenting Eko: Production of Space in a Nigerian City', in Anthony King (ed.), *Re-Presenting the City: Ethnicity, Capital and Culture in the 21st Century Metropolis* (London: McMillan, 1996), pp. 111–136.

[14]To mention a prominent few: Otonba Payne, *Table of Principle Events in Yoruba History* (Lagos: Andrew Thomas, 1893); Obadiah Johnson, *Lagos Past* (Lagos: no pub, 1901); Samuel Johnson (ed.), *The History of the Yorubas: from the Earliest Times to the Beginning of the British Protectorate* (London: George Routledge, 1921), John Losi, *History of Lagos* (Lagos: African Education Press, 1967 [1914]).

[15]Losi, *History of Lagos*, p. 1.

[16]Law, 'Trade and Politics', p. 344.

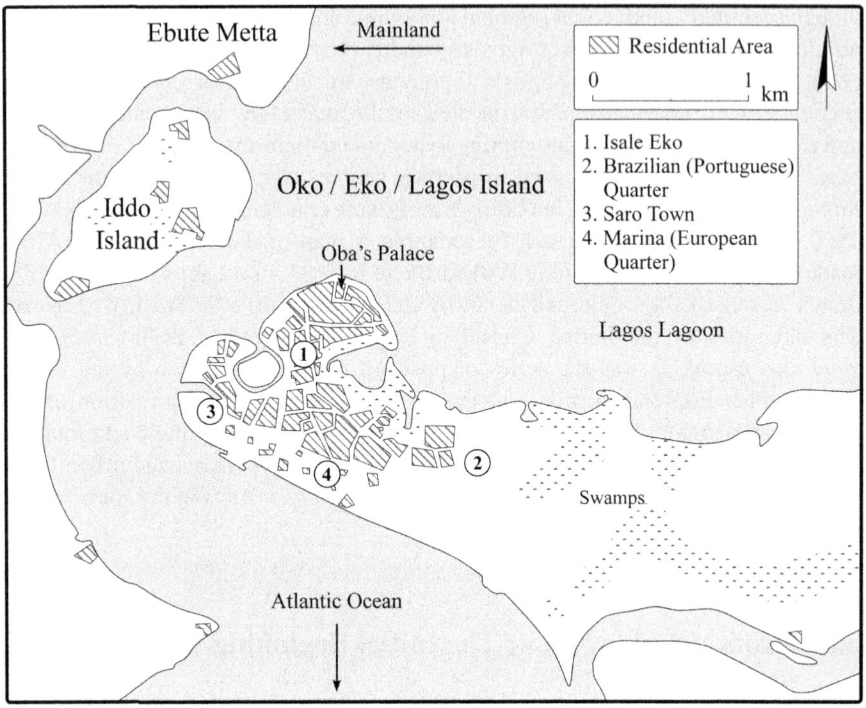

Fig. 5.1 The four main quarters in Lagos in the 1880s (Map drawn by the authors, based on the British Archives, CO 700/ Lagos 5, plan of the town of Lagos in 1883)

attributed by recent historians to the fifteenth century: the move southwards to the fertile land of Ebute Metta, still part of the mainland, proceeding beyond Iddo Island, as Ebute Metta was vulnerable to attack by neighbouring groups. Though well-positioned for defence, Iddo Island –one square mile of sandy land in all – could hardly provide for the food needs of its occupants, so further land on a southerly island beyond the lagoon was sought, from which the first appellation of Lagos – *Oko*, meaning 'a farm' in Yoruba – was derived[17] (Fig. 5.1).

The thriving Yoruba urban tradition since medieval times recognised a clear division between the central space, designated for residence (*Ile*, i.e., 'a home') and its outskirts, beyond the city's encircling mud walls, which were normally used as farmland (*Oko*, i.e., 'a farm'). Yet in Lagos from the start this *Ile* vs. *Oko* dichotomy was not apparent, with *Oko*, the initial settlement, considered as *ile*, and its morphology including farms that gradually became residential land or mixed-use. The physical structure of the four-square-mile Lagos Island and the vagueness of traditional spatial terminology clarifies these morphological particularities.

[17] Aderibigbe, 'Early History of Lagos', pp. 3–4.

Until the mid-nineteenth century the town occupied no more than a third of the overall surface of the island. A thick encircling belt of mangrove forest acted as a barrier against rip-tide silt as well as human penetration, and the interior was mostly swamps, creeks and dry land. Passage between the mainland and the island was therefore chiefly by canoe. The even coastline of the Island today is a result of swamp reclamation by the British colonial regime from the 1860s.[18] The highest point of Oko was at its extreme northwest, where one of the first founding fathers chose to erect his red pepper farm, a site now occupied by the palace of the *Oba* (king) of Lagos. The pepper farm became the centrepiece of a new prosperous village inhabited by fishermen and farmers, and ever since its establishment on the farm site, the palace has been known as *Iga Idunganran*, that is, 'the pepper palace.'[19]

Which building materials where identified then in *Oko* and its composing residential units? Later condemned as insanitary by the colonial regime, these were temporary, most probably lagoon mud, palm leaves, poles, bamboo, decayed vegetables and clay (no physical evidence has survived). One can assume that those dwellings followed the Yoruba compound model of the family residential unit. Prosperous houses in Yoruba towns were mainly rectangular, with rooms built around an open courtyard, the number and size of rooms depended on the family composition, social status of the head, and household hierarchy. The apartment of the head of household was for his exclusive use and faced the entrance, while a separate room was allocated both to each of his wives and her children, other rooms being allocated to relatives and domestic slaves.[20]

The change of Lagos's name from Yoruba *Oko*, 'a farm', to Bini *Eko*, 'a military camp', is related to the presence of the eastern Benin kingdom in the area, especially from the mid-seventeenth century. Lagos was a vassal state of Benin, paid tribute, and introduced Bini's advisors and rites to its court. Its dynastic *Oba*s remained Yoruba, yet Benin was the source of their royal legitimacy, and the Bini connection lasted until the establishment of British influence in Lagos in the 1850s, even while Benin power declined over the eighteenth century, and Lagos grew commercially and militarily independent.[21]

Bini residential patterns were introduced, with the influence of Ile-Ife not the only model, but it may be a little strained to say that "these [Bini houses in Lagos] were built close to one another and arranged in good order. They had roofs, verandahs and balustrades and were covered with leaves of palm trees and banana."[22] Similarly, it is impossible to reconstruct details such as carved doors in a virtually Bini style assigned to contemporary Lagos (Fig. 5.2) – these are only assumptions, based on

---

[18]Bigon, *A History of Urban Planning*, pp. 40–44.

[19]W. H. Irvine, *The Iga Idunganran* (Lagos: Federal Government Printing, 1966); Akinsemoyin and Vaughan-Richards, *Building Lagos*, p. 6.

[20]Michael Barbour (et al.), *Nigeria in Maps* (London: Hodder & Stoughton, 1982), pp. 98–99.

[21]Law, 'Trade and Politics', pp. 327–333.

[22]Akinsemoyin and Vaughan-Richards, *Building Lagos*, p. 9.

Fig. 5.2 Reconstruction of a Yoruba household from the times of the original settlement (Akinsemoyin and Vaughan-Richards, *Building Lagos*, p. 8)

early European descriptions of Bini houses in the city of Benin itself.[23] Nothing is known of the plan of the first Oba palace, said to have been influenced by the palace of Benin.[24] The resemblance in pronunciation between the Yoruba word *Oko* and its Bini counterpart *Eko* may have facilitated the adoption, common usage, and eventual replacement of one toponym by the other. The northwestern part of the Island, on the slope of the Oba palace, became the prominent residential quarter, known as *Isale Eko* ('Lower Lagos'), now considered as central or downtown Lagos.[25] As mentioned before, the name Eko has been preserved in the indigenous culture as the main local appellation for Lagos.

5.3 From 'Curamo' to 'Onim': Portuguese Imprints

Another cultural influence, which affected the island's spatial organisation, during the reign of Oba Akinsemoyin (second half of the eighteenth century), was that

[23]Such as the Dutch report from the sixteenth century cited, without a reference, in:
Alain Vaughan-Richards, 'Le Nigeria', in Jacques Soulillou (ed.), *Rives colonials* (Paris: ORSTOM, 1993), pp. 243–288 (p. 245).

[24]Vaughan-Richards, 'Le Nigeria', p. 248.

[25]Aspects in this continuity were broken with the evacuation of parts of this quarter in a slum clearance scheme during the 1950s decolonisation. National Archives Nigeria (Ibadan) (hereafter NAI) I, Com. Col. I, 3860/ I.

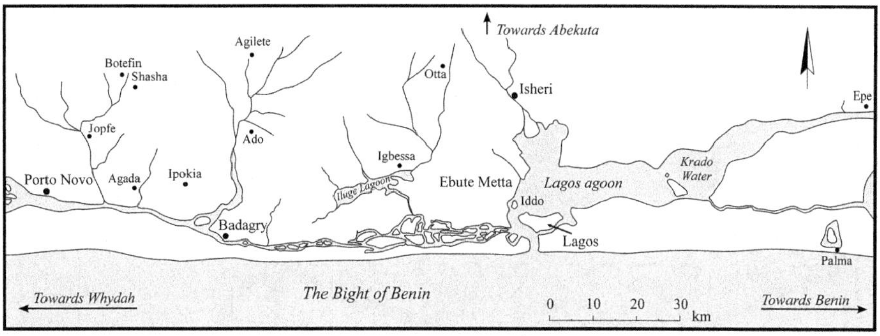

Fig. 5.3 The Lagoon network connected Porto-Novo with Benin, through Lagos Lagoon (Redrawn by authors after the original at the British Archives, MR 1/145, Lagos and southern Nigeria, 1893)

of Portuguese slave dealers who were reportedly invited to Eko. Akinsemoyin met them in Whydah – today's Ouidah in Benin (country) to the west, not Benin city to the east – during his earlier exile during succession conflicts in Eko.[26] Whydah, along with other coastal towns such as Porto Novo and Badagry, had traded for a long time with Portuguese and other European merchants and slavers, and indigenous towns were competing for foreign attention, entangling hinterland political entities in a complex network of changing alliances and rivalries (Fig. 5.3). Eko integrated quite late into this network, with direct European trade apparently starting in the 1760s as part of a movement of European interest eastwards from Whydah in reaction against rigid Dahomian trade policies.

At this time further names were added to the indigenous names of Lagos, though the local name *Eko* is not usually mentioned in European sources of the time. A map by Dutch cartographer Willem Blaeu (published in 1635) gave the name *Ichoo* to an imaginarily extended island along the Bight of Benin (Fig. 5.4). Robin Law identifies this with the indigenous name *Eko*, and attributes the cartographic inaccuracy to the map-maker's reliance upon fragmentary verbal reports of the topography. Maps of the time also showed wild animals, both realistic and imaginary, in order to fill an empty space yet to be 'discovered' and measured. One of the main Portuguese names for Lagos and its environs during the seventeenth century, Curamo, probably originated in the Bini word *Eko*. Oral traditions of African groups of this region (Ijebu, Itsekiri and Ijo) refer to Lagos as *Korame*, *Ikurame* or *Ukuroama;* these groups had direct or indirect contact with Portuguese and other European traders who may have preserved one version or another of the Bini phrase *Eko ne ame* ('Eko [military camp] on the river').[27]

---

[26]Aderibigbe, 'Early History of Lagos', p.11.

[27]For a reference to Blaeu's map of Guinea and for the identification of the Bini phrase and other regional traditions we are indebted to Robin Law. See especially the very detailed notes, numbers 44 and 46, in his aforementioned 1983 article.

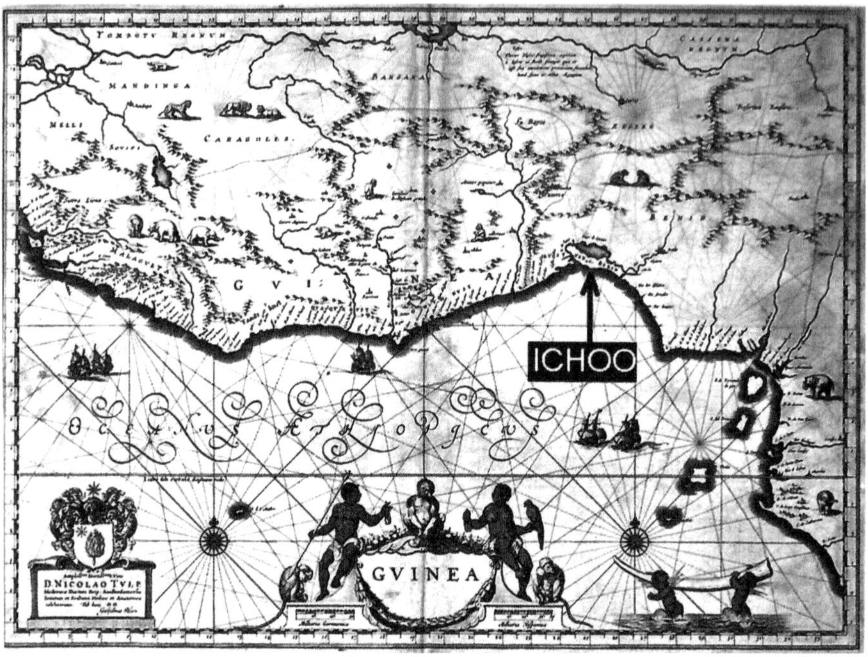

Fig. 5.4 Blaeu's map of West Africa, Amsterdam 1635, with 'Ichoo' for 'Lagos' (Added by the authors; reproduction, private collection)

During the seventeenth and early eighteenth centuries, Lagos and sometimes its neighbouring environment were referred to as *Curamo* in Portuguese sources and in those of other European trading powers (such as Dutch, French and British) active along the West African coast. Similar variations were apparent for Lagos and its environs, particularly on maps, such as *Lago do Curamo* (i.e., 'Lake of Curamo'), *Caran* or *Karam* and *Kuramo*, while the Lagosians were sometimes referred to as *Couranas*. The toponym *Curamo* was used in parallel to another Portuguese name that became more dominant in the late eighteenth and early nineteenth centuries – *Onim;* variations in contemporary sources, especially Portuguese and French, were *Aunis*, *Ahoni* or *Onis*. These names, as explained by Law, probably derived from *Awori*, the Yoruba sub-group to which the first residents of *Oko* belonged.[28] Pierre Verger noted that between the mid-eighteenth and mid-nineteenth centuries Lagos maintained established connections with the Bahia kingdom in Brazil through Lisbon, and contemporary documents from Bahia call Lagos 'Onim.'[29]

The toponym *Onim* became more prevalent from the 1760s, when direct trade began between Lagos and the Europeans with the invitation to the Portuguese by

[28]Law, 'Trade and Politics', pp. 330–331, 342–343.

[29]Verger, 'Notes on some Documents', pp. 344, 350.

Oba Akinsemoyin. Portuguese traders were given land around Campos Square, an area to the southwest of Isale Eko which became known as the 'Portuguese Town', and later as the Brazilian Quarter (Fig. 5.1). The whitewashed houses of the Portuguese, reminiscent of their metropolitan provincial cities, were aligned along streets with trees, and were simple, one-storied buildings. The Portuguese gave precious presents in return for a trading monopoly – such as deep-jet-satin velvet that caused Akinsemoyin's palace hall, according to Losi, to look dark at daytime and bright at night – and also rebuilt Akinsemoyin's palace with materials shipped from Portugal. *Iga Idunganran* was roofed with iron columns and clay tiles, a real innovation which saved the palace from at least two conflagrations,[30] and remained an outstanding feature of Lagos building until the 1850s. As Consul Scala observed in his *Memoirs*, the *Iga Idunganran* was: "the biggest of all [. . .] distinguished from others by its roof which is made of tiles instead of leaves, a privilege to which he [the Oba] alone had entitled."[31] Under British influence, clay tiles were changed to corrugated-iron roofing, but other Portuguese influences survived in the palace until the 1960s, when Oba Adele II asked for architectural assistance to enlarge the palace. The new addition in cement was marked with a publication on the history of the building, including photos of the older rooms and courtyards, which showed clear Portuguese influence in the internal arcades, baroque-style balustrades, supported pillars and capitols.[32]

Lagos was thus already a polyglot or hybrid town on the eve of the British intervention in 1851 and their establishment of a consular regime. Lagos's markets, owing their prosperity to the slave trade and to products such as cloth, food and palm oil, attracted many immigrants from various parts of Nigeria-to-be and beyond. In its rich ethnic composition, Lagos was thus distinct from other pre-colonial Yoruba settlements, and was not a monocultural native town. The hub of the area was still the *Iga Idunganran* of Isale Eko, the first and most important quarter in Lagos. The aristocracy of Awori Yoruba and Bini was gradually joined by other Yoruba groups, each settled in their own area.

5.4 From 'Onim' to 'Lagos': On the Eve of the Colonial Period

In 1901 the Yoruba historian Obadiah Johnson concluded that two main events affected contemporary Lagos. The first was the Fulani invasion from the north and the internal Yoruba wars, the second the advent of the British missionary, trader and administrator, while the connection between the two was, according to him,

---

[30]Losi, *History of Lagos*, pp. 13–14.

[31]Robert Smith (ed.), *Memoirs of Giambattista Scala, Consul of His Italian Majesty In Lagos in Guinea, 1862*, trans. Brenda Packman (Oxford: Oxford University Press, 2000).

[32]W. H. Irvine, *The Iga Idunganran* (Lagos: Federal Government Printing, 1966).

"the foreign slave trade."[33] Pressures by Fulani Islamic groups from Hausaland southwards brought about the collapse of the Old Oyo Empire of the Yoruba around 1835. The alliances between various Yoruba powers that comprised this empire, already weakened, were now broken, as these powers invaded each other's territories and demolished each other's towns, generating a flow of refugees into new Yoruba settlements. In addition, arms and ammunition originally from the coast were brought to the hinterland by missionaries, while the only stable factor remained the movement of slaves to the coast.[34] Lagos became a chief slave trade centre at the expense of Porto-Novo, where French slavers were held back by the Anglo-French war in 1793 and the abolition of the slave trade in France a year later. British slavers still operated in Lagos, with over 7,000 slaves exported annually until 1800[35] – a sufficient supply during the Oyo wars. Even the subsequent British abolition of slavery had little effect on Lagos's position as the only place along the western coast where the slave trade was still in operation, chiefly due to its physical environment, which increased the operation there of vessels of the British Royal Naval Squadron. Lagos Lagoon was unapproachable for large vessels not only because its mouth was shallow and sometimes blocked by dangerous swampy mud, but because of a network of coastal lagoons, parallel to the Atlantic but unreachable and hidden from the ocean, which allowed human access by canoes.[36] This network, facing the Bight of Benin yet unseen from it, interconnected several coastal towns stretching from Porto Novo towards Benin, with Lagos in the middle (Fig. 5.3).

The persistence of the slave trade in Lagos Island led to the British bombardment in 1851, initiating the British 'Consular Period' that lasted for 10 years, during which time the British Consul, previously based offshore at Fernando Po, was transferred to Lagos. The former Oba was restored, following his consent to sign a treaty that prohibited slave trade and human sacrifice and protected missionaries and free trade. Contemporary British accounts presented moral disapproval of slavery as the main reason for their occupation of the area[37]; whereas other European scholars prefer to emphasise the economic factor in the British seizure.[38] Nigerian historians have identified political rivalries in Lagos and adjoining territories, each trying to

[33]Johnson, *Lagos Past*, p. 27.

[34]Jean Herskovits 'Liberated Africans and the History of Lagos Colony to 1886' (unpublished PhD thesis, St. Hilda's College, Oxford, 1960), pp. 38–46.

[35]Law, 'Trade and Politics', p. 347.

[36]The British Archives, MR 1/145, Lagos and southern Nigeria, 1893.

[37]Leighton Wilson, *The Destruction of Lagos* (London: Ridgway, 1852), pp. 1–23; H. Straith, *The Yoruba Mission: Peace-Making Intervention of the Missionaries at Ado and the Affair at Lagos* (London: Seeleys, 1853), pp. 12–16; William Nevill Geary, 'The Development of Lagos in 50 Years', *West Africa*, 16 (1924), n.p.; Margery Perham, *Native Administration in Nigeria* (Oxford: Oxford University Press, 1937), p. 13.

[38]Especially Anthony Hopkins, 'Property Rights and Empire Building: Britain's Annexation of Lagos, 1861', *Journal of Economic History*, 40, 4 (1980), pp. 777–798.

take advantage of the circumstances and acting in self-interest, with the British getting increasingly involved.[39]

The name *Lagos* itself – from the Portuguese *lago* or *lagõa* ('Lake' or 'lagoon') – replaced the other names only when Portuguese influence started to fade and gave way to the British. Existing before in various versions such as *Rio de Lagos*, *Rio de Lagua* or *Rade de Lagos* (which means 'Lagos River' or 'Anchorage'), these versions also suggest that West African Lagos might have taken the name of the eponymous port in southern Portugal, from the mid-fifteenth century the main centre of Portuguese expeditions down the African coast, introducing through its slave markets the first African slaves into post-medieval Europe.[40]

Between 1854 and 1856, with the consolidation of British control, the name *Lagos* finally replaced *Onim* and the latter disappeared from European sources. This distinction was made by Pierre Verger, with no further explanation.[41] The toponym 'Lagos' has persisted from the British colonial period to the present, and the continuity between the Portuguese toponym and the political events and building forms was apparently broken, but our research shows the *persistence* of the toponym, political change and built form.

5.5 Between 'Lagos' and 'Eko': The Colonial Period

The Consular Period marked the establishment of a *pax Britannica* both among the opposing regional factors and between them and the British, a peace further consolidated by the formal annexation of Lagos in 1861. News about the 1851 bombardment of Lagos and the Oba's restoration attracted to Lagos new groups of immigrants, repatriates and expatriates. These heterogeneous inflowing groups reflected Lagos's physical and strategic situation as a crossroad, "a point where highways from the interior, by rivers, lagoons and through the forest, meet the highways of the ocean"[42] (Fig. 5.3). In the course of this process, Lagos was practically organised in the form of four quarters (Fig. 5.1), as following: the first was Isale Eko, consisting of 30,000 residents in 1859.[43] Though its street layout was considered 'chaotic' in the eyes of the colonisers, who fought the use of temporary building materials by legal means – this quarter was considered as the source of indigenous political legitimisation by the Yoruba. As the base of the

[39]Obaro Ikime, *The Fall of Nigeria: The British Conquest* (London: Heinemann, 1977); J.F. Ade-Ajayi, 'West Africa in the Anti-Slave Trade Era', in John Fage and Roland Oliver (eds), *The Cambridge History of Africa*, vol. 5 (8 vols.) (Cambridge: Cambridge University Press, 1976), pp. 200–222; A conversation between Prof. Ade-Ajayi and the author, Ibadan, 2007.

[40]Malyn Newitt, *A History of Portuguese Expansion: 1400–1668* (London: Routledge, 2004), pp. 24–28.

[41]Verger, 'Notes on some Documents', p. 350.

[42]Robert Smith, *The Lagos Consulate 1851–1861* (London: Macmillan Press, 1978), p. 128.

[43]Baker, *Urbanization and Political Change*, p. 24.

Fig. 5.5 'Brazilian'-style private residence dated to the late-nineteenth century, Lagos (Photo by Liora Bigon)

establishing forefathers and the Obaship – the latter institution controlled then on land allocation – the name *Eko* always simultaneously preserved in everyday local use.

The second quarter was the Portuguese Town, renamed the Brazilian Quarter. The 'Brazilians' were self-emancipated slaves who settled down there in the 1880s, as the Portuguese slave traders were displaced by British 'legitimate trade.' This quarter's residents – recorded as 3,000 in 1881, about 9 % of the total population[44] – brought from Bahia and Brazil valuable skills, particularly in building. Better materials allowed the addition of a second and even a third storey to buildings (Fig. 5.5), outstanding in the contemporary single-storey townscape.

The third quarter was the Saro Town, inhabited by former slaves, immigrants from Sierra Leone (duly called *Saro*-s), many emancipated in the Atlantic by the British Royal Naval Squadron. After the British seizure of Lagos, their immigration grew to 2,500 by the 1860s, about 10 % of the total population.[45] Living in the south-western part of town as an educated elite, the Saros cooperated with the British, financing the grand residences built by the 'Brazilians.'

It might be an overstatement to call the few grouped warehouses of the European traders after 1851 as the fourth or 'European Quarter', but that thin strip along the southern shore of Lagos Island would become within the next decades the

[44]L. C. Dioka, *Lagos and Its Environs* (Lagos: First Academic Publishers, 2001), p. 181.

[45]Baker, *Urbanization and Political Change*, p. 21.

Marina – a desirable and prestigious part of Lagos for the colonisers. For the colonised, however, it was considered as the 'back yard' of the town, serving as burial land for socially excluded persons and what was believed as their evil spirits. Such contradicted conceptions, with their economic and socio-political differentiation, co-existed throughout the colonial period, making for confusion and confrontation.[46] The number of Europeans in Lagos was not great, reaching 250 by the end of the nineteenth century.[47] Apart from the European residences and business, colonial administrative buildings and the house of the British governor served as a focal point for land transportation from the hinterland and international maritime transportation.

What was therefore the toponymic importance of the quarter-like structure of Lagos, including the meaning of the persistence of the name 'Lagos' during the British colonial regime and afterwards?

The Lagos Treaty of Cession of 6 August 1861 marked the formal initiation of British rule in Lagos Island, and also Eurocentric cartography and history. From this point, the Portuguese name *Lagos* became the only official name for *Eko*, although the indigenous population and the African repatriate resident groups continued to refer to Lagos Island as *Eko*. The name *Lagos* was internationally accepted by other European agencies active there (as testified by Consul Scala[48]), and was the only name used by the British colonisers and European cartographers in identifying the area on world maps.

The persistence of the name seems to have related to the character of British colonial rule in West Africa, and its philosophical justification as a 'civilising' mission. Lagos was known, at least until the discovery of the cause of malaria in 1900, as 'the white man's grave', and British 'indirect rule' policy sought colonial economic development with as little interference as possible with indigenous society.[49] This doctrine was associated especially with Lugard, Governor-General of the amalgamated Nigeria, who detested Lagos and avoided staying there.[50] The West African colonies were expected to support themselves financially, so that a mere 400 Europeans administered the territory of Nigeria which is several times larger than the United Kingdom, with one result that all the existing toponymy not only stayed as was, but was rather also embraced by the colonisers, including the Yoruba names of almost all Lagos's quarters, sub-areas, streets and thoroughfares.[51]

---

[46]Geary, 'The Development of Lagos', n.p.; Hopkins, 'Property Rights and Empire Building'; Peter Marris, *Family and Social Change in an African City: A Study of Rehousing in Lagos* (London: Routledge, 1961), ch. 1.

[47]Akin Mabogunje, *Urbanization in Nigeria* (London: University of London Press, 1968), p. 242.

[48]Smith, *Memoirs of Giambattista Scala*.

[49]Frederick Lugard, *The Dual Mandate in British Tropical Africa* (London: Blackwood, 1922).

[50]Frederick Lugard, *Lugard and the Amalgamation of Nigeria: A Documentary Record*, compiled and introduced by A.H.M. Kirk- Greene (London: Frank Cass, 1968), p. 90.

[51]This is clearly understood from NAI, Com. Col. I, 2148, Place names, Lagos Municipal Area, 1938–55.

Nigeria, unlike French Algeria, Southern Rhodesia (Zimbabwe) or South Africa, was never a white settler colony, and thus its urban space was neither dramatically transformed nor subject to major place naming or renaming by the coloniser. The adjacent island to Lagos (today part of Metropolitan Lagos) was renamed Victoria Island following major swamp reclamation and the cutting of a new canal by the British, and the larger territory of the British crown colony of Lagos was named 'Colony Province' after the amalgamation of Southern and Northern Nigeria in 1914. In Eastern Nigeria the new port created close to strategic coal deposits was named Port Harcourt by Lugard after the British Colonial Secretary of the time. But these were exceptions. Other than within the areas reserved for Europeans (European Residential Areas), the British in Nigeria were reluctant or hesitant to impose European place names. The various administrative headquarters in Northern Nigeria kept the Hausa-Fulani names of their former indigenous settlements (Lokoja, Zungeru, Kaduna), and even for the new layouts near the railway it was the African names that stuck. Thus in Ibadan, the trading area next to the railway, intended to be known as 'Lebanon Street' for the Lebanese who traded from there, the name that persisted was 'Gbagi' (after the survey pegs' that marked the plots, in Yoruba[52]). In Northern Nigeria 'native reservations' for Lugard's category of 'native foreigners' were known from the start of Sabon Gari (or 'strangers' quarter), the pre-colonial indigenous name for such districts, while 'Tudun Wada' in Zaria, a mixed residential and trading layout, meant a place for making money.

The Lagos map of 1885, at a scale of 1: 3,240, produced as a basis for land registration and tax collection, and prepared by a Saro land surveyor, documented and accepted the Yoruba toponymy of the Island in detail, so that it was absorbed into the official repertoire of the colonial state.[53] It included, for instance, 'Ehin Igbeti' (the 'cursed land') for the Marina, and 'Tinubu Square' (named for Madame Tinubu who had resisted British rule and been exiled), later renamed 'Independence Square.' The rare series of the French photographer Edmond Fortier, images taken in the 1900s, includes an image of 'Docemo street', commemorating Oba Dosunmu who in 1861 was obliged to sign the treaty bringing the Island under British control (the British could hardly pronounce his name, which was adapted as 'Docemo') (Fig. 5.6).

---

[52]Such plot marking system, a symbol of European geographical violence and political dominance, yield resistance in other colonial contexts: in Waitara and Ngatimaru (New Zealand) in 1860 and 1872, the survey pegs were pulled out by the Maori, challenging the highly contested land purchase. Giselle Byrnes, 'Surveying – The Maori and the Land: An Essay in Historical Representation', *New Zealand Journal of History*, 31, 1 (1997), pp. 85–98 (p. 93).

[53]Cambridge University Library, Map Collection: Plan of the Town of Lagos, West Coast of Africa, prepared by W.T.C. Lawson, 30 December 1885, 1:3,240. Maps. AA.583.88.1 (is not incorporated due to the issue of publishing rights).

Fig. 5.6 Fortier's postcard showing 'Docemo' street, Isale Eko, Lagos (Private collection)

The boundaries of Lagos colony were enlarged to include, by the late nineteenth century, the adjoining Ikoyi and Victoria Islands, the offshore Iddo Island and part of Ebute-Metta on the mainland,[54] and subsequently further mainland areas were annexed, creating eventually an immense urban sprawl with the ever-expanding fringe areas (Fig. 5.7). Around the time of independence in 1960 there were about 400,000 residents in Lagos Island and immediate environs, and Lagos soon after became the first West African city to pass the million mark.[55] With about ten million residents today, it is the fastest growing city in Africa, competing only with Cairo.

5.6 A Post-Colonial Glimpse

Since independence changing the country's name or the name of its then capital has not been debated in Nigeria, as far as we are aware, unlike in other African countries, where an extensive toponymic 'Africanisation' occurred after independence, especially in former white settler colonies. The relatively low friction between

[54]O. Adejuyigbe, 'Evolution of the Boundaries of Lagos', *Nigeria Magazine*, July/Sep (1969), pp. 480 484.

[55]P. O. Sada and A. Adefolalu, 'Urbanisation and Problems of Urban Development', in A. B. Aderibigbe (ed.), *Lagos: The Development of an African City* (London: Longman, 1975), pp. 79–105.

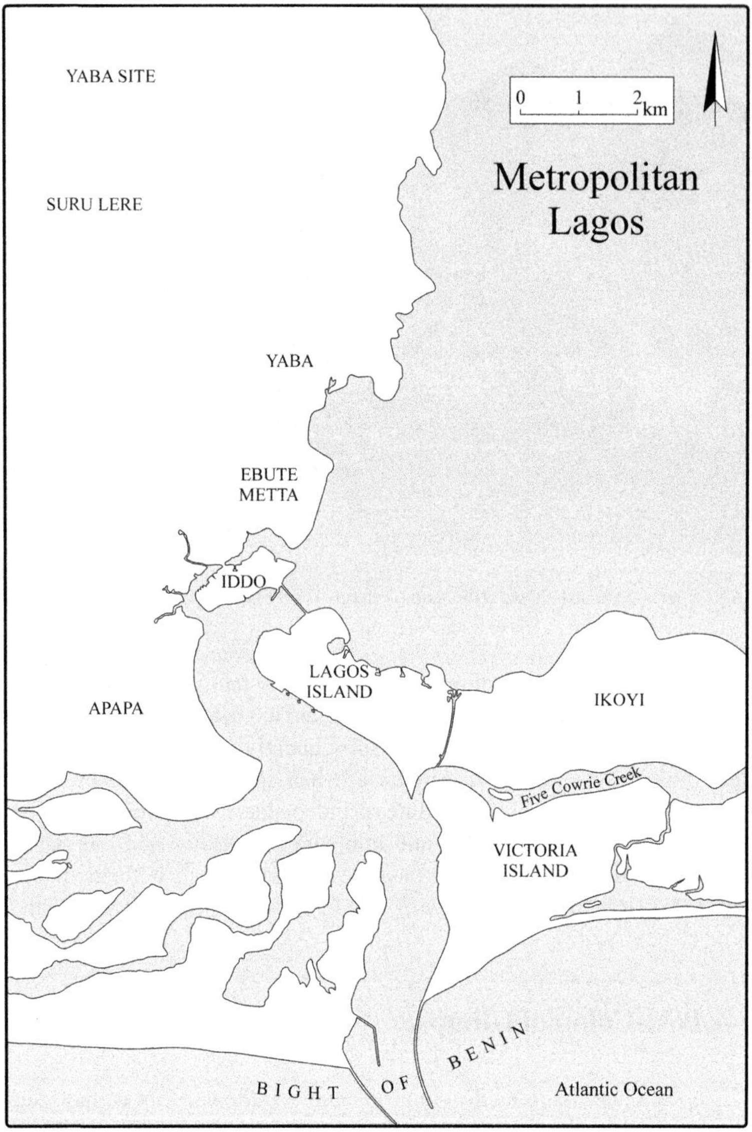

Fig. 5.7 Lagos Metropolitan area in the 1940s and 1950s (Map drawn by the authors)

the coloniser and the colonised in Nigeria included a situation where most of the indigenous typonomy had been accepted by the colonial regime, hence little need for a radical toponymic de-colonisation. Lagos's street names did, however, become objects for frequent renaming in disputes between Nigeria's federal government and

Fig. 5.8 Statute of the Idejo chiefs, Lagos (Photo by Liora Bigon)

the local municipal authorities. The street names were subjected to internal splits between political parties within a year of independence.[56]

Another assumption as to the persistence of the Portuguese toponym 'Lagos' concerns its 'indigenity.' Though the Portuguese period included uncomfortable memories such as the slave trade, the Portuguese preceded the other colonial forces in their trading contacts with West Africa after 1445, imparting longevity to their toponymy, so that the name 'Lagos' is not necessarily identified with the colonial period, and hence has not needed de-colonising.

Standing at the entrance to the city on the Lagos-Ibadan-Benin highway, a modern cement statute of three of the Idejo chiefs represents the pre-colonial authorities of Lagos (Fig. 5.8). The Idejo or 'white-cap chiefs' were appointed by the Oba to allocate lands, one of the most sensitive and complicated issues in the colonial period and afterwards. The statue not only converses with the inclusive and multi-layered chronological approach to African cities suggested by Coquery-Vidrovitch, but, through Lagosians' use of the name 'Eko', also statue

[56]G. Dawodu, 'Naming of Streets in Lagos', *The Service*, 25 March, 1961.

symbolises the tough struggle of living in the mega-city, considered as vivid, dangerous and chaotic. According to popular interpretation, each of the three Idejos says respectively: 'you shall not be stupid, you shall not be slow, you shall not allow yourself to be taken for a fool in Eko.'[57]

[57] In Yoruba: 'O gbodo ridin, O gbodo suegbe, O gbodo ya mugum l'Eko' (authors' conversation with Lagosians, Lagos, 2007).

Chapter 6
From 'Avenue de France' to 'Boulevard Hassan II': Toponymic Inscription and the Construction of Nationhood in Fes, Morocco

Samira Hassa

Abstract Using four maps of the city of Fes (1908, 1933, 1953, and 1986) this chapter examines the evolution of the toponymy of Fes, a post-colonial city, across three historical eras: the pre-colonial period, the French occupation, and 30 years after the end of the French protectorate. Fes is split into two well-marked districts: The Ville Nouvelle or 'New City', built by the French; and the Medina, the old town, with its Muslim traditions and Arab urban architecture. While the toponymy of the Medina did not change, the Ville Nouvelle underwent changes in street naming that reflect changing political and social conditions. The 1933 map shows many street names honouring French military heroes, while the 1953 map shows more street names commemorating French intellectuals and an increase in Arabic street names, suggesting that the colonial French power was fading. In contrast, the 1986 map indicates that the toponymy gradually switched from French to Arabic in the Ville Nouvelle as result of the Moroccan nationalist movement. These changes of denomination are a symbolic reclamation of the previously colonised space and an affirmation of Moroccan-Muslim identity.

Keywords (Post-) colonial Morocco • Fes • Cartography • French urbanism • North African city • Muslim heritage • Arab/French toponyms

6.1 Introduction

A street name identifies and demarks but also acts as a symbolic sign of the political and cultural power in play. In a colonial context, architecture and toponymy are called upon to testify to the authority of the coloniser, providing a primary playground for the politics of space and place underlining strategies of exploitation,

S. Hassa (✉)
Modern Languages and Literatures, Manhattan College, New York, NY, USA
e-mail: samira.hassa@manhattan.edu

L. Bigon (eds.), *Place Names in Africa*, DOI 10.1007/978-3-319-32485-2_6

domination, and exclusion.[1] This chapter analyses street naming changes and the motives behind those changes in the city of Fes, Morocco. Although Rabat is the official capital, Fes is considered the cultural and spiritual capital of Morocco. During the French protectorate (1912–1956), Moroccans retained partial control of political and cultural affairs while the French controlled the economy and foreign affairs. We shall therefore expand on the toponymy of Fes to understand it as a historical narrative of French political ideology and a reflection of how Moroccans reacted to the dominance of the coloniser at the end of the French occupation in 1956. As Anthony King writes, 'Toponymic inscription suggests clues about the different categories – social, cultural and spatial – into which units of urban space have been divided. Particular names, their etymologies and histories, or their distribution across space, open windows on the past, providing information about the spatial distribution of social or political power.'[2] In this chapter, four maps of Fes, produced during different eras – pre-colonial, colonial, and thirty years after the end of the French protectorate – that reveal these changing social and political dynamics will be examined: a 1908 map published in the periodical *Le Temps* on June 15, 1908; a 1933 map produced by the US Army Map Service; a 1953 map produced by the Institut Géographique National, Annexe du Maroc, Rabat; and a 1986 map produced by the Division de la Cartographie de Rabat. Only the 1908 map is shown below because it is in the public domain, while the other maps are copyrighted. Examples of toponyms from the other maps are listed in Tables 6.1, 6.2, and 6.3.

My approach is 'bottom up', starting with the maps themselves to understand the French colonial authority and the Moroccans' counter-reaction to it at the end of the French occupation. Photos taken in Fes during ethnographic fieldwork in 2015 illustrate the data. Given that the maps are from different time periods and made by different agencies, some have more toponymic information than others. However, comparing the street names on each map shows street-naming patterns and gives some clues as to colonial ideologies as well as post-colonial changes.

The written language found in public spaces such as street-name signs, graffiti, and advertising is receiving special attention from historians, sociologists, and urban planners as well as linguists as this discourse carries evidence of social, cultural, and political interactions and conflicts. Extensive studies have investigated street-naming decisions as a historical narrative. Duncan Light, Ion Nicolae, and Bogdan Suditu examined changes in street naming in Bucharest from 1948 to 1965 of streets commemorating personalities of the Soviet and Romanian communist eras

[1] This sentence draws on the following sources, though the colonial reality is astoundingly missing from the second: Henri Lefebvre, *The Production of Space*, trans. Donald Nicholson-Smith (Oxford: Blackwell, 1991); Michel Foucault, 'The Language of Space', in Stuart Elden and Jeremy Crampton (eds), *Space, Knowledge and Power: Foucault and Geography* (Burlington, VT: Ashgate, 2007), pp. 163–168.

[2] Anthony King, *Spaces of Global Cultures: Architecture, Urbanism, Identity* (New York: Routledge, 2004), p. 142.

Table 6.1 Some examples of street names on the 1933 map of Fes in the Ville Nouvelle

Place/ Square	Avenues/ Boulevards	Rues/Streets	
Place Gambetta	Avenue de France	Rue des Etats Unis	Rue Clemenceau
Place Lyautey	Avenue du Général Maurial	Rue de Belgique	Rue Jean Jaurès
Place Galliéni		Rue Indochine	Rue Roland Fréjus
Place Lafayette	Avenue Foch	Rue du Japon	Rue Pierre Loti
Square Le Guevel	Boulevard de Verdun	Rue de Serbie	Rue Montaigne
	Boulevard du 4ème Tirailleurs	Rue D'Angleterre	Rue Alfred de Vigny
	Avenue des Sports	Rue d'Italie	Rue Pasteur
	Avenue de la Gare	Rue de Pologne	Rue Lavoisier
	Avenue de Meknes	Rue du Portugal	Rue Coccard
	Avenue de Sefrou	Rue de Russie	Rue du Dr Auvert
	Boulevard Moulay Youssef	Rue d'Espagne	Rue des Jardins
		Rue de Tunisie	Rue du Ravin
	Route de Fes à Sefrou	Rue d'Algérie	Rue du 11 Novembre
		Rue de Savoie	Rue du Général Gouraud
		Rue d'Auvergne	Rue du Capitaine Cuny
		Rue de Picardie	Rue du Commandant Prokos
		Rue de Provence	
		Rue de Normandie	Rue Charles de Foucault
		Rue de Bourgogne	Rue Léon L'Africain
		Rue Bringau	
		Rue Samuel Biarnay	

Table 6.2 Some examples of street names on the 1953 map of Fes in the Ville Nouvelle section

Place/ Square	Avenues/ Boulevards	Streets
Place de l'Atlas	Avenue du Général De Gaulle	Rue Molière
Avenue des Almohades	Boulevard du Général Ducla	Rue Malherbe
Avenue Louis Barthou	Avenue du Général Duc Colombat	Rue Marcel Cerdan
	Avenue Lartigue	Rue Avicennes
	Avenue des Almohades	Rue Averroes
	Avenue de Monastir	Rue Bergounier
		Rue René Caillié
		Rue Victor Hugo

as a clear manifestation of communist power.[3] In Dakar and Lagos, Liora Bigon examined urban planning and colonial doctrines as manifestations of assimilation and indirect rule, arguing that in Dakar, French policy excluded the natives' identity and toponyms referring to natives, while in Lagos the opposite was observed as

[3]Duncan Light, Ion Nicolae and Bogdan Suditu, 'Toponymy and the Communist City: Street Names in Bucharest, 1948–1965', *GeoJournal*, 56 (2002), pp. 135–144.

Table 6.3 Some examples of streets renamed on the 1986 map of Fes in the Ville Nouvelle

Protectorat	Post-Protectorate
Place Gambetta	Place de la Résistance
Place Galliéni	Place Ahmed el Mansour (Sultan of Morocco 1578–1603)
Place Lyautey	Place de Florence
Boulevard Poeymirau	Avenue Mohamed V (King of Morocco 1927–1961)
Avenue Général de Gaulle	Avenue Youssef Ben Tachaffine (Sultan of Morocco 1061–1106)
Avenue du Général Colombat	Avenue Allal Ben Abdallah (Moroccan anti-colonialist)
Avenue du Général Maurial	Avenue Mohamed Ezzarktouni (Moroccan anti-colonialist leader)
Avenue Louis Barthou	Avenue Abou Obaida Ben Jerrah (Companion of the Prophet Mohamed and commander of the Muslim army)
Avenue Foch	Avenue Farhat Hachad (Tunisian union leader and anti-colonialist)
Rue de Belgique	Rue du Liban
Rue de Russie	Rue de Syrie
Rue de Savoie	Rue Al Hoceima
Rue Bringau	Rue Patrice Lumumba (Congolese independence leader)

a result of a more laissez-faire British colonial policy.[4] In the context of colonial Singapore, Brenda Yeoh indicates that two-thirds of commemorative street names honoured European figures while the rest referred to Asians in order to display inclusion of Asians as racial segregation was illegal in Singapore.[5] Focusing his research on commemorative street naming, Maoz Azaryahu notes that in the city of Berlin after 1945, East Germany cleansed itself of reminders of Nazism in street naming and in 1974 18 streets were renamed to honour anti-Nazi martyrs.[6] Despite this growing interest in street-name choices and the motives for them, research has been lacking on the topic in Morocco. As noted in the introduction of the present collection, there is a great need for critical studies of toponymy in Africa.

6.2 The City of Fes and French Colonisation

Like many post-colonial cities in the Southern Hemisphere, Fes is divided into two main parts: the indigenous (pre-colonial) section and the ex-colonial section that

[4]Liora Bigon, 'Urban Planning, Colonial Doctrines and Street Naming in French Dakar and British Lagos, c. 1850–1930', *Urban History*, 36, 3 (2009), pp. 426–448.

[5]Brenda Yeoh, 'Street Names in Colonial Singapore', *Geographical Review*, 82 (1992), pp. 313–322.

[6]Maoz Azaryahu, 'Street Names and Political Identity: The Case of East Berlin', *Journal of Contemporary History*, 21, 4 (1986), pp. 581–604.

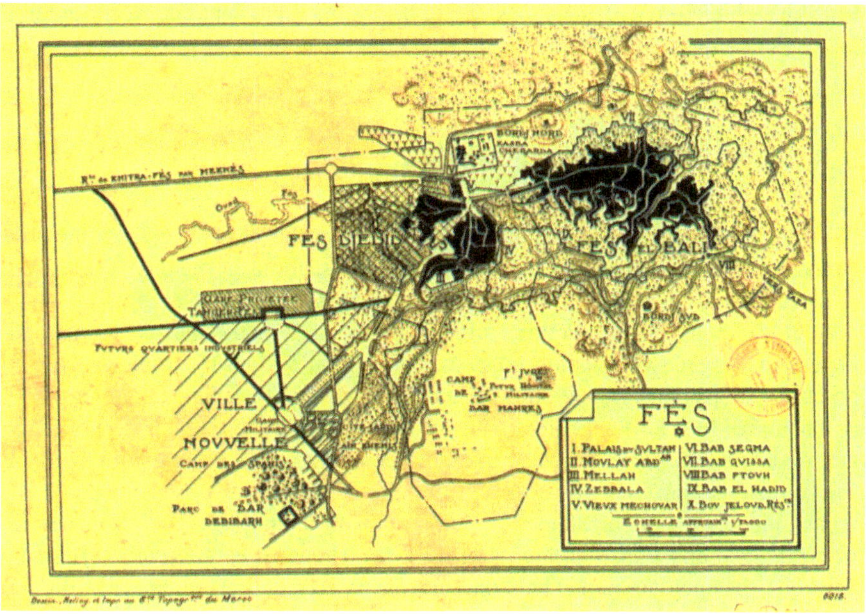

Fig. 6.1 Map of Fes, dated 1918 (Courtesy of Gallica.bnf.fr/ Bibliothèque nationale de France)

was designated for European expatriates (Fig. 6.1). Fes el Bali, literally 'Old Fes' in Moroccan Arabic, more commonly called the Medina, is considered the heart of the city. This section was built in 789 by the great grandson of the prophet Moulay Idris I. Adjacent Fes El Jdid, 'New Fes', was built in the thirteenth century. While those two sections of Fes have Arabic names, the third part of the city has a French name: Ville Nouvelle, 'New City.' This section – a generic phenomenon in French North Africa – was built with the arrival of the French after 1912. The division between the Ville Nouvelle and the Medina reflects the racial segregation that aimed to separate the French from the Moroccans, the coloniser from the colonised, the modern educated European from the uneducated indigenous Moroccan.[7] Janet Abu-Lughod explains that urban division in Morocco implemented under Hubert Lyautey (Resident-General from 1912 to 1925) virtually took place for sanitary purposes due to fear of disease and even compares it to an urban apartheid.[8] This urban and architectural dichotomy between the Ville Nouvelle and the Medina is striking. The Ville Nouvelle is largely influenced by Parisian styles, with broad avenues, angled buildings, balconies, and terraces while the Medina illustrates traditional Arabic and

[7]Gwendolyn Wright, 'Tradition in the Service of Modernity: Architecture and Urbanism in French Colonial Policy, 1900–1930', *Journal of Modern History*, 59 (1987), pp. 291–316.

[8]Janet Abu-Lughod, *Rabat: Urban Apartheid in Morocco* (Princeton, NJ: Princeton University Press, 1980).

Muslim architecture with its arches, fountains, zellige terra cotta tile work, and tight passages ending in cul-de-sacs.

6.3 Street Naming in the Medina

A visitor to Fes immediately notices how different the Medina is from the Ville Nouvelle, since a thousand years separate the two sections of the city. The Medina is listed as a UNESCO world heritage site as it is considered an open medieval museum with its Islamic architecture. Walking through the narrow, curved and zigzagging streets while being pushed aside to make room for donkeys carrying goods for shops (cars are not allowed here), the Medina feels like a labyrinth. Even as a native of Fes, I often get lost. The pre-colonial 1908 map (Fig. 6.2) shows only the Medina as the Ville Nouvelle had not been built yet. Toponyms are relatively scarce on this map. Several start with the word *bab* ('door') as the Medina is a typical medieval city surrounded by ramparts with main gates. Two examples are *Bab Khoukha* and *Bab Boujloud.* The map also shows *Souk Al Kamis* ('Thursday Market'), the open-air market held on Thursday in preparation for the Friday Jumma

Fig. 6.2 Map of the Médina of Fes, 1908 (courtesy of Gallica.bnf.fr/ Bibliothèque nationale de France)

prayer, followed by a shared meal, usually with couscous. Other toponyms on the 1908 map are *Kaisariya* ('mall'), *Karaouin*, the Islamic university and mosque founded in the ninth century, and *Moulay Idriss*, the mausoleum, considered the holiest place in Fes. There are also more descriptive toponyms, like *Kantra Touila*, the 'long bridge', and *Kantra Kebira*, the 'big bridge' in Moroccan Arabic.

The 1933 and 1953 maps provide more information about street names during the protectorate. On the 1933 map, more toponyms use the word *bab*, as in *Bab Ftouh* and *Bab El Hedid*. The main axis of the Medina is designated *Talaa Kbira*, 'large uptown street.' Some street names pertain to traditional Moroccan handcrafts: *Nejjarine* ('wood'), *Chrablliine* (traditional Moroccan shoes), and *Fekharine* ('ceramics'). Toponyms honoring personalities are also present, for example *Dar Tazi*, referring to the home of Pasha Tazi, the mayor of Fes. On the 1933 map the toponym *Résidence* designates the Dar Menebhi, the palace that served as Lyautey's first home. Lyautey formed an alliance with the elite merchants of Fes in keeping with France's collaborative political tactic designed to increase French domination[9]; this may explain why he chose to live in the Medina.

On the 1953 map, schools and health infrastructures are more visible, suggesting that a significant number of health and educational services were purposely placed in the natives' section of the city during the French occupation, as they were intended to facilitate a pacifying relationship between the French and Moroccan civilians. The map shows many *écoles* ('schools'), such as *Ecole Française de Bou JeLoud*, *Ecole de l'Adaoua*, and *Ecole des Notables Indigènes* – the latter was a school for the native upper class as a strategy to create an alliance between the Moroccan social elite and the French. Educating Moroccans, especially the sons of the elite. was part of Lyautey's plan to cultivate a relationship between the French and the natives. Several health and charity facilities are shown on the 1953 map (*hôpital Murat*) along with numerous medical dispensaries (*Dispensaire Al Adaoua*, *Dispensaire de Bab Ftouh*) and an orphanage called simply *Orphelinat*. Public health and politics worked hand in hand in the colonial era, when health was an important aspect of the civilising mission, projecting an image of the coloniser's humanism. This humanism, as reported by Louise Willis, fascinated the American author Edith Wharton (1862-1937) during her visit to Morocco as she was impressed by Madame Lyautey's charitable efforts, especially the maternity hospitals called *La goutte de lait* ('drop of milk') that she created in the medinas of several Moroccan cities.[10]

The newer 1986 map shows many more toponyms, such as *Dar Al Batha and Lycée* ('secondary school') *Oum Al Banine*, to help visitors navigate the old town, but there are no major changes, including to street names, from the previous maps here in the Medina. Lyautey intentionally avoided changing the Medina and in

[9]Moshe Gershovich, *French Military Rule in Morocco: Colonialism and Its Consequences* (London: Frank Cass, 2000).

[10]Edith Wharton, 'Madame Lyautey's Charitable Works in Morocco', trans. Louise Wills, *Tulsa Studies in Women's Literature* 17, 1 (1998), pp. 29–36.

this section of Fes street names were not affected by the French occupation. All toponyms in the Medina from 1908 to 1986 were specific to the Moroccan Arab culture, with no reference to French culture. This was in line with the association doctrine, a policy based on collaboration and cooperation between the French and the indigenous population designed to limit Moroccan anti-colonial resistance and to strengthen French colonial power.[11] This also reflects Lyautey's goals to preserve the Medina's authenticity and even "impose a romanticized vision of the medina as a medieval relic",[12] what Paul Rabinow calls 'techno-cosmopolitanism'[13]; that is, a strategy that aims to bring modernisation while protecting traditional local culture in order to maintain social control – despite the desire of local people to modernise the Medina.[14]

The toponyms shown in the Medina on the maps from the different periods are not especially precise; they seem to act more like landmarks. This mode of reference is widespread in many urban contexts in Africa and beyond.[15] As Michel Ben Arrous also reminds us in this collection, the notion of unambiguous toponymic inscription was alien to pre-colonial entities, and African societies are not always identified with affixed toponyms or even located where the toponyms say they are. This imprecision is perhaps the case in the Moroccan culture, traditionally based on oral communication as business deals and marriages were transacted only orally. This is connected in turn to the fact that Moroccan Arabic is usually a spoken language while the written form is reserved for standard Arabic – the language of the Qur'an and the language learned in schools. The landmark tradition still exists – if you take a cab today in Morocco or give directions, people may not know the name of the street but will rely rather on a nearby landmark to find it. Today authorities are making efforts to add street names and signs in the Medina, often written in Arabic, to provide distinct addresses (Fig. 6.3). In some places, urban planners have chosen street names and in other cases they have opted for street numbers, perhaps because they are easily remembered and understood by those who are not literate.

---

[11]Raymond F. Betts, *Assimilation and Association in French Colonial Theory, 1890–1914* (New York: Columbia University Press, 1960).

[12]Stacy E. Holden, 'Constructing an Archival Cityscape: Local Views of Colonial Urbanism in the French Protectorate of Morocco', *History in Africa*, 34 (2007), pp. 121–132 (p. 124).

[13]Paul Rabinow, 'France in Morocco: Technocosmopolitanism and Middling Modernism', *Assemblage*, 17 (1992), pp. 52–57 (p. 53).

[14]Stacy E. Holden, 'When it Pays to Be Medieval: Historic Preservation as a Colonial Policy in the Medina of Fez, 1912–1932', *Journal of The Historical Society*, 6, 2 (2006), pp. 297–316.

[15]Yeoh, 'Street Names'; Ambe J. Njoh, 'Toponymic Inscription, Physical Addressing and the Challenge of Urban Management in an Era of Globalization in Cameroon', *Habitat International*, 34 (2010), pp. 427–435; Belkacem Boumedini and Nebia Dadoua Hadria, 'Les noms des quartiers dans la ville d'Oran: entre changement officiel et nostalgie populaire', *Droit et Cultures*, 64, 2 (2012), pp. 149–160.

Fig. 6.3 Street sign of Rue Arsat Bennis in the Médina (Photo by Samira Hassa)

Fig. 6.4 Street sign of Rue du Commandant Mellier, renamed Rue Ahmed Amine (Photo by Samira Hassa)

6.4 Street Naming in the Ville Nouvelle

Compared to the Medina, it is immediately noticed on all the maps that in the Ville Nouvelle streets are clearly designated and there are more toponyms (Table 6.1). On the 1933 map, corresponding to the first decades of the French protectorate, there is a preponderance of streets honouring military commanders, with street names displaying military ranks such as *Général, Commandant, Capitaine*, or *Lieutenant* (Fig. 6.4). These street names make the power structure present and visible to sustain organisation and domination, exemplifying what Michel Foucault calls *pouvoir-*

savoir or power-knowledge.[16] Among the streets honouring military commanders are the Avenue du Général Maurial, for the commander of the Fes region, and Avenue Général Foch, for one of the most celebrated French military theorists of the early twentieth century. Other military personalities, perhaps less important, were honoured on smaller streets, such as the Rue du Général Gouraud and Rue du Colonel de Castries. The strong presence of military figures on the maps is not surprising given the French involvement in World Wars I and II, when military commanders were considered heroes. The wars themselves were commemorated in the streets of Fes named after major battles, as in the Boulevard de Verdun, after the 1916 battle, and the Rue du 11 Novembre, the date of the 1918 armistice ending World War I.

Other Ville Nouvelle streets on the 1933 map honour countries, including France, Italy, Portugal, Spain, and Russia. The Avenue de France designates the largest and main avenue of the city of Fes, strongly denoting French domination. Germany, France's enemy during this period, is notably absent. Some streets were named after countries that were under French control, including the Rue d'Indochine, Rue D'Algérie, and Rue de Tunisie, emphasising the extent of the French Empire and its power. The regions of France were not forgotten: on the 1933 map of Fes one can find the Rue de Provence and Rue de Picardie, bringing some local French flavour to Morocco and connecting the colony to France. The greatest toponymical honour that could be bestowed on an individual was to name a plaza after him. Joseph Galliéni (1849–1916), military commander in chief of the colonies, under whom Lyautey served in Algeria and Indochina,[17] was given a plaza, as was Lyautey. Léon Gambetta (1838–1882), a French statesman and one of the founders of the Third Republic, was given a much more prestigious spot as the Place Gambetta was directly connected to the Avenue de France. More republican figures were represented in the physical space of Fes, including France's leader during World War I, Georges Clemenceau (1841–1929), imposing the republican ideology of the Third Republic, an era during which French imperialism was rationalised through the humanistic rhetoric of the 'civilising mission.'[18] The French socialist leader Jean Jaurès (1859–1914) was also honoured with a street, but it was far from the city centre and the Avenue de France, perhaps due to his anti-colonial views.[19] Other French administrators who served in Morocco were also remembered, such as Le Guevel, director of municipal services from 1922 to 1925, who was honoured with the Square Le Guevel, and Samuel Biarney, a speaker of Arabic and Berber who

---

[16]Michel Foucault, *Power/knowledge: Selected Interviews and Other Writings, 1972–1977* (New York: Pantheon, 1980).

[17]Moshe Gershovich, 'The Ait Ya'qub Incident and the Crisis of French Military Policy in Morocco', *Journal of Military History*, 62 (1998), pp. 57–73.

[18]Alice Conklin, 'Colonialism and Human Rights, a Contradiction in Terms? The Case of France and West Africa, 1895–1914', *American Historical Review*, 103 (1998), pp. 419–442.

[19]Jacques Marseille, 'La Gauche, la Droite et le fait colonial en France. Des années 1880 aux années 1960', *Vingtième siècle. Revue d'histoire*, 24 (1989), pp. 17–28.

perhaps because of his linguistic skills was appointed to lead the Islamic religious foundation called Habous.

Other streets commemorated travellers with a connection to Morocco, such as seventeenth-century French traveller Roland Fréjus, Charles de Foucault (1858–1916, and Léon L'Africain (approximately 1488–1554); and nineteenth-century French novelist Pierre Loti, author of *Au Maroc* (In Morocco), among other works, as well as French literary canons such as Montaigne and Alfred de Vigny. There are very few references to Moroccan culture on the 1933 map. It shows the Boulevard Moulay Youssef connecting the Avenue de France with the Medina. Moulay Youssef was the conciliatory brother who served as sultan of Morocco from 1912 until his death in 1927 after his brother, Sultan Moulay Hafid, was forced to abdicate by the French. Placing the Boulevard Moulay Youssef strategically between the Ville Nouvelle and the Medina was an obvious sign of inclusion of the local power. Some toponyms (Table 6.1) are more general and do not carry politically loaded content. They are either directional or descriptive, as in the Avenue des Sports, Avenue de la Gare (Train Station Avenue), Rue des Jardins (Gardens Street), Rue du Ravin (Ravine Street), and Route de Fes à Sefrou (Road from Fes to Sefrou, a city 20 miles south of Fes).

As the Ville Nouvelle expanded, new streets appeared on the 1953 map (Table 6.2), including some named after public figures such as General de Gaulle, who was given an avenue, and Louis Barthou, the French prime minister and foreign minister who was assassinated along with Yugoslavian King Alexander in 1934. On the 1953 map there are more streets referring to literary, art, and sports personalities, such as the Rue Molière and Rue Marcel Cerdan, nicknamed *le Bombardier marocain* (the Moroccan Bomber), a French boxer who was born in Algeria but grew up in Morocco. There are also more toponyms referring to Islamic scholars such as Avicennes (980–1037) and Averroes (1126–1198), as well as toponyms referring to the history and geography of Morocco, as in the Place de l'Atlas and the Avenue des Almohades, commemorating the twelfth-century Almoravide Dynasty. The addition of toponyms related to Morocco and Muslim culture on the 1953 map shows perhaps that the French were making concessions, hinting that the colonial French power was beginning to waver.

The French protectorate in Morocco impacted not only the physical space but also education, banks, hospitals, and of course linguistic issues. From 1912 to 1956 French became the language of education, business, and the media. At independence in 1956, the question of the position of the French language within the already rich linguistic landscape of standard Arabic, Moroccan Arabic, and the different varieties of Berber spoken in Morocco needed to be dealt with. The Istiqlal ('independence' in Arabic) nationalist party issued the Arabisation language policy to restore the prestige of Arabic, the language of the Qur'an, and limit the use of French, the language of the coloniser.[20] As the Moroccan socio-linguist Ahmed Boukous

[20]Gilbert Grandguillaume, *Arabisation et politique linguistique au Maghreb* (Paris: Maisonneuve, Larose, 1983).

explains, "The aim of the arabisation policy is to impose standard Arabic as the language of operation in all public institutions, ideally occupying all domains of communication and monopolising both oral and written language, as French is a serious competitor in the domain of written communication."[21] This policy was also imposed on the language of public spaces, including street signs, to mark a new linguistic, cultural, and political regime that displayed the Moroccan Arab identity and attempted to eliminate the French presence (Table 6.3).

The 1986 map shows streets honouring post-independence Moroccan nationalist leaders such as Allal Ben Abdallah, leader of the anti-colonisation movement (in 1953 he attempted to assassinate Sultan Mohammed Ben Aarafa, who was chosen by the French to replace Sultan Mohamed V, whom they had exiled to Madagascar); Muhammad Allal al-Fassi, founder of the Istiqlal nationalist party; and Tunisian union leader and anti-colonialist Farhat Hachad, killed in 1952 by the French secret service. Another street commemorates Patrice Lumumba, Congolese anti-colonial leader and prime minister killed in 1961.[22]

These toponymic inscription choices show solidarity with the anti-colonial movement at large and reinforce the toponymic cleansing plan to erase the French colonial authority from the Moroccan space. The largest avenue of Fes, the Avenue de France, was replaced by the Avenue Hassan II, after the king of Morocco from 1961 until his death in 1999. The Boulevard Poeymirau, named after Lyautey's military and administrative subordinate[23] was changed to the Avenue Mohamed V, after the king of Morocco from 1927 to 1961. The marocanisation and nationalisation of the toponymy of the Ville Nouvelle continued with renaming the Place Gambetta as the Place de la Resistance, a name with clear anti-French colonial intentions. Place Lyautey was changed to Place Florence as Fes and Florence are sister cities, suggesting that western names and cultures were not the problem, but those connected to France. French names needed to be banished from the Moroccan space to show re-appropriation by Moroccans of their territory (Table 6.3). Anthony King reports a similar phenomenon in New Delhi, India, where streets were changed from English to Hindi names commemorating national leaders following independence from Great Britain.[24]

Streets named after regions of France on the 1933 and 1953 maps disappeared, to be replaced by Moroccan cities and regions, such as the Rue Tetouan, Rue du Rif, Rue Marmoucha, Rue Tafilalet, and Rue Al Akandar, all regions largely inhabited by Berbers. Streets named after Arab cities and countries display Morocco's belonging to the Arab world, as in the Rue du Caire and Rue de Syrie. In addition, some dates

---

[21] Ahmed Boukous, 'Globalization and Sociolinguistic Stratification in North Africa: The Case of Morocco', in Cécile Vigouroux and Salikoko Mufwene (eds), *Globalization and Language Vitality: Perspectives from Africa* (New York: Continuum, 2008), pp. 126–141 (p. 131).

[22] In 2002 the Belgian government admitted partial responsibility and issued an apology for his death: 'Apology for Lumumba Killing', *New York Times*, 6 February 2002.

[23] Gershovich, 'The Ait Ya'qub Incident.'

[24] Anthony King, *Spaces of Global Cultures: Architecture, Urbanism, Identity* (New York: Routledge, 2004), p. 146.

were easy to change: Rue du 11 Novembre was changed to Rue du 16 Novembre, corresponding to the return in 1955 of Sultan Mohamed V, grandfather of the current king Mohamed VI, from exile in Madagascar.

While many streets were renamed, some kept their original names. For obvious reasons, Boulevard Moulay Youssef survived, along with non-political descriptive street names such as Avenue du Ravin and Avenue des Sports. Surprisingly, some streets named for countries, like Rue des Etats-Unis and Rue du Canada, keep their original names while others – perhaps perceived as neutral new western countries that could offer political and economic opportunities to Morocco, such as Rue de Serbie and Rue d'Indochine – disappeared.

6.5 Conclusion

This chapter has examined toponymic inscriptions longitudinally using original artefacts to allow maps to unfold their version of the French colonial story in Morocco. The maps show that the toponymy of the Medina did not change over time as it was anchored in local features, giving the illusion of French respect for Moroccan culture and traditions. In contrast, on the 1933 and 1953 maps of the Ville Nouvelle the preponderance of street names are related to French culture and history, attesting to the expansion of French proprietorship and exclusion of Moroccan culture from the physical space. As shown by the post-colonial map, at independence Moroccans responded to the dominance of France by purging street names from the French colonial regime and renaming streets after anti-colonial figures and references to the Moroccan Arab culture, revealing the public space as a site of performance of power. Today, as one walks through the streets of the Ville Nouvelle, a street sign from the French colonial regime occasionally appears (e.g., Fig. 6.4) as those are only slowly being removed, a reminder that the colonial past is not so past.

Chapter 7
The Colonial Toponymic Model in the Capital Cities of French West Africa

Hélène d'Almeida-Topor

Abstract Old urban areas or centres, developed during the colonial period in French-speaking sub-Saharan Africa, are characterised by specific toponymy, largely exogenous, as a considerable part is inherited from the colonial period. The naming of roads and streets in the colonial urban environment was carried out by an official act or more often by a decree from the colonial Governor during a management-board meeting and published in the *Journal Officiel* of the territory. A systematic analysis of the minutes of the period, which we undertook for each territory of French West Africa (AOF), and also the analysis of the minutes of appropriate sessions of management boards – enable us to reconstitute the toponymic history of the capital cities of the AOF. This 'top-down' study of place-naming policy is intended to analyse the situation towards the end of the colonial period, that is, the period of decolonisation when the countries became independent, and had the choice to accept or reject these marks of forced integration into an empire.

Keywords French West Africa • Colonial capitals • Colonial urban toponymy • Street-names typology • *Annuaire Vert/Guide bleau AOF*

Old urban areas or centres, developed during the colonial period in French-speaking sub-Saharan Africa, are characterised by specific toponymy, largely exogenous, as a considerable part is inherited from the colonial period.[1] The naming of roads and streets in the colonial urban environment was carried out by an official act or more often by a decree from the colonial Governor during a management-board meeting and published in the *Journal Officiel* of the territory. A systematic analysis of the

[1]This contribution was originally published in French under a similar title in *La ville européenne outre mers: un modèle conquérant (15e-20e siècles)*, edited by Catherine Coquery-Vidrovitch and Odile Goerg (Paris: l'Harmattan, 1996). Reprinted by permission of behalf of l'Harmattan. English translation, minor stylistic changes and further annotations were made by the Editor of this volume.

H. d'Almeida-Topor (✉)
Centre de recherches africaines, Université Paris I – Sorbonne, 152 rue de Tolbiac,
75013 Paris, France
e-mail: dalmeida-topor.helene@neuf.fr

L. Bigon (eds.), *Place Names in Africa*, DOI 10.1007/978-3-319-32485-2_7

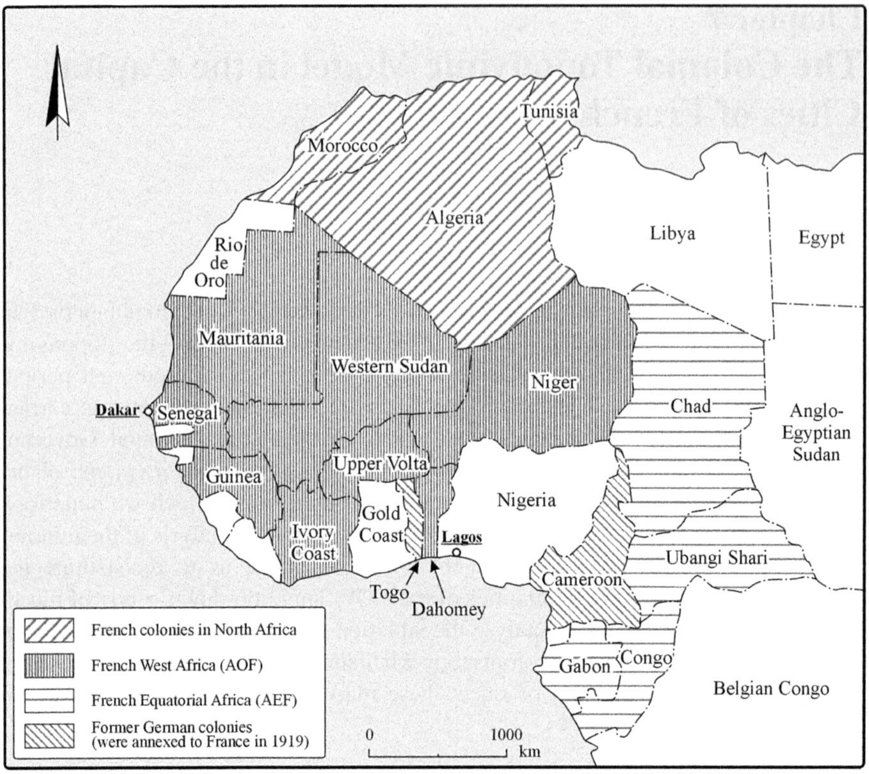

Fig. 7.1 Colonial West Africa (Map drawn by Liora Bigon)

minutes of the period, which we undertook for each territory of French West Africa (AOF)[2] (Fig. 7.1), and also the analysis of the minutes of appropriate sessions of management boards – enable us to reconstitute the toponymic history of the capital cities of the AOF. This 'top-down' study of place-naming policy is intended to analyse the situation towards the end of the colonial period, that is, the period of decolonisation when the countries became independent, and had the choice to accept or reject these marks of forced integration into an empire.

[2]The AOF (*Afrique Occidentale Française*), that is the federation of French West Africa, was created in 1895, alongside the neighbouring, later, federation of French Equatorial Africa (AEF). The aim was to facilitate the centralist decision-making process in Paris. The AOF's overall territory amounted to 4,633, 985 km^2, and included eight colonies: Senegal, French Sudan (today's Mali), French Guinea, Ivory Coast, Dahomey (today's Benin), Upper Volta (today's Burkina Faso), Niger and Mauritania. Following the First World War, Togo, a Mandate territory, was included as well. Jean Suret-Canale, *French Colonialism in Tropical Africa, 1900–1945* (London: Hurst & Co., 1971), p. 308 [the Editor].

7.1 An Exogenous Model

Though the toponymy of the pre-colonial towns has not been a subject of systematic research, and the diversity of these urban areas urges caution before being a subject of any generalisation,[3] the studies that are available show that spatial identification was generally based on quarters, squares and the markets.[4] The names of these places were derived from geographic criteria or historic circumstances. As the spatial growth of the capitals of the pre-colonial states demonstrates, the quarters were normally named with the proper name of a King or prince who founded them, or after an event. For example, such as in the case of Abomey: the word 'Gbégamé', which in Fon means 'in the high grass', has became the quarter of 'Djegbé.' The later name, given following the installation of the future King Glèlè there, is a contraction of 'you are the salt which no language can refuse.'[5] The other quarters of Abomey took the surnames of the lineages who were settled in them, while the squares took the name of the patron deity or of the ceremonies which took place there periodically.[6] With regards to the main roads and thoroughfares, in entire regions they were named principally after their geographical location or the directions to which they would lead.

The old indigenous modes and practices of place naming were continued until after the French conquest and were even adopted in urban quarters that were created during the colonial period. Moreover, the growth of the cities was marked by the development of new neighbourhoods, many of which were designated by the nationality or the ethnic origin of the migrant communities who were settled there. For instance, the *zongo* – that is the 'strangers' quarter' that is also called 'Sabon Gari' in West Africa – constitutes a pre-colonial element that continued into the colonial period and even perpetuated under the French and British colonial regimes. Present in towns throughout the coastal areas, it brought together the migrants from the Sahel region.[7] However, the impact of colonisation resulted, inter alia,

[3]Catherine Coquery-Vidrovitch, *Histoire des villes d'Afrique noire:* des *origines à la colonisation* (Paris: Albin Michel, 1989), ch. 1.

[4]Jacques Soulillou (ed.), *Rives colonials: architectures, de Saint-Louis à Douala* (Paris: Parenthèses/ORSTOM, 1993).

[5]The capital of the kingdom of Dahomey, which is situated in present-day Benin and whose official language is Fon – the same as the name of the population that founded it. S. Agninikin, *Etude sur l'évolution historique sociale et spatiale de la ville d'Abomey* (Cotonou: URBANOR, 1986), p. 13. See also: Michael Houseman, Blandine Legonou, Christiane Massy and Xavier Crepin, 'Note sur la structure évolutive d'une ville historique: l'exemple d'Abomey (République populaire du Bénin), *Cahiers d'études africaines*, 26, 104 (1986), pp. 527–546.

[6]Agninikin, *Etude sur l'évolution historique* , p. 19.

[7]In Hausa, 'Zongo' signifies 'a camp', with temporary or permanent housing. See Rose Koupaki, 'Histoire de la ville de Cotonou de 1920 à 1945', PhD dissertation (Dakar: Université de Dakar, 1986), which includes an in-depth study of the toponymy of Cotonou's quarters. With the establishment of the colonial regime, migrants – such as railway workers – were also displaced from the coastal regions into *Zongo*-s within the northern Sahelian towns [the Editor].

in specific identification of urban spaces, based on roads and streets. The latter were usually organised in a grid or orthogonal pattern, which is common with minor modifications in most capitals of French West Africa. The main parallel roads – called boulevards or avenues as applicable – are intersected, often at right angles, by other avenues or important streets, creating a network of relatively large roads.

The subdivision into lots, which constituted the privileged mode of occupying the soil reserved for future extension of the colonial cities, used predetermined numbers, as was the case for roads meant for movement. The latter were subsequently named, according to the circumstances. In some cases, as is the case of Treichville, a quarter of Abidjan, these numbers were partially given with names; in others, as is the case of the Médina quarter of Dakar, numbers were unique. At the same time, the roads of commercial and residential areas of the French colonial capitals, including big business centres, were quickly named during particular occasions or events, such as visits by notable individuals, commemorations, etc. For example, in Dahomey (present day Benin) a decree dated 1912 awarded the name 'Boulevard Henri Malan' to "the avenue which goes down from Avenue Ballot at the lagoon along the new building of the customs house."[8] This was enacted during the meeting of the Board of the Administrative Council that took place on 19 July as homage to the former governor of Dahomey who had died a few days earlier in Indo-China.

By the end of the 1950s, the roads and streets in the capitals of the AOF were almost all named. While these names were more or less used by the urban residents, it was generally in a selective way – a phenomenon whose reasons require a detailed comparative study. Similarly, in Conakry, though all the roads were officially named, the residents continued to use a semi-official numbered system in their reference to boulevards and avenues.[9] One of the reasons for this practice might be that in all the urban areas, the postal address was reduced to the post-box number (B.P.), what did not lead itself to the utilisation of names in everyday practice. The advertisement of various businesses, as appeared in the *Annuaire Vert* (1955), helps us to understand the differences between one city and the other. For instance, the company Travaux-Afrique, with its headquarters in Paris, was located at Dakar '28 rue Jules Ferry, BP 735'; at Conakry '6 boulevard BP 294'; and at Lomé 'BP 289.' Thus, the colonial toponymic model has entered into use in a limited and uneven fashion. It constituted,

[8] *Journal Officiel du Dahomey*, decree dated 19 August, 1912 (p. 270).

[9] *Annuaire Vert,* 1955. Notice that in Conakry there are three layers of naming systems that are currently in use in the ex-colonial city centre (Kaloum): the colonial system, comprised of proper names; the post-colonial system, which embraced numerical system for most of its orthogonal streets, leaving the colonial names of some main boulevards only; and the World Bank's Urban Development Program which introduced, in the 1990s, a new numerical system throughout the grid network. The following part of the paragraph refers indeed to a fourth method, partly formal, based on postboxes [the Editor].

and still constitutes in the urban centres of the former AOF cities, an official façade, and also appears in the official plans of these urban areas.[10]

7.2 Colonialism in Daily Life

We recorded the names assigned to main urban roads in order to develop a typology and to make a comparison mainly between the capital cities of French West Africa. The analysis was performed on 336 names, though the corpus obtained is not comprehensive and the process of listing was somewhat exhausting. This is because by the end of the 1950s, there was no complete inventory of the main roads except for Dakar, as that of Abidjan was carried out much later.[11] It was therefore necessary to cross-check the information provided in the phone books and carry out a study of city plans, particularly those of the *Guide bleu*. This guide was first published in French West Africa in the year 1958.[12] It partially explains the significant disparities in the inputs taken into account, nevertheless the main reasons are the disparity in areas and the density of urban populations.

Concerning Table 7.1 below and the corpus of toponymic inscriptions examined, with regard to Dahomey (present day Benin) we chose to include Cotonou in our

Table 7.1 Composition of the corpus examined

	Number of roads classified	Percentage
Dakar	109	32.4
Bamako	54	16.0
Abidjan	45	13.4
Bobo-Dioulasso	36	10.7
Conakry	34	10.1
Lomé	18	5.4
Porto-Novo	16	4.8
Cotonou	14	4.2
Ouagadougou	10	3.0
TOTAL	336	100

Sources: *Annuaire Vert* (1955), *Guide bleu, AOF* (1958)

[10]This is clear from an on site experience versus consulting the series of plans of these formerly colonial cities, with differing importance, held by the Department of Maps and Plans of the French National Library.

[11]For Dakar see: *Annuaire Vert*, 1955; For Abidjan see: *Guide d'Abidjan*, no date (published in the early 1960s). File classified Gc FF 17479, French National Library.

[12]Houlet Gilbert, *Les Guides bleus: Afrique occidentale française*, first edition (Paris: Édition Hachette, 1958).

Table 7.2 Typology of the toponymic themes in the end of the 1950s

	Location:			Personalities with local influence:	
	France/world	Africa	Activities	French	African
	(A)	(B)	(C)	(D)	(E)
Abidjan	22.2	11.1	15.6	51.1	–
Bamako	14.8	27.8	18.5	35.2	3.7
Conakry	23.5	23.5	5.9	41.2	5.9
Cotonou	28.6	7.1	14.3	50	–
Porto-Novo	25	18.8	6.3	50	–
Dakar	26.6	32.	5.5	32.1	3.7
Lomé	55.6	16.7	5.6	22.2	–
Ouagadougou	30	–	10	40	20
Bobo-Dioulasso	11.1	22.2	25	30.6	11.1

Sources: *Annuaire Vert* (1955), *Guide bleu, AOF* (1958)

study due to its economic role, although the official capital was Porto-Novo (also included). For a similar reason, for Upper Volta (present day Burkina Faso) we took into account Ouagadougou, the official capital, as well as Bobo-Dioulasso. On the other hand, it was not possible for us, due to lack of information, to carry out a comparative study for Nouakchott (Mauritania) and Niamey (Niger). In addition, except for Dakar, the rest of the number of roads identified is estimation.

The criteria employed for the typology are aimed to evaluate the colonial effects on the urban toponymy (see Table 7.2). Under the first category the names linked to a geographical location or a particular activity were grouped together. For example, in almost all the cities in question there were names that referred to the local landscape and physical features, whether natural or man-made, such as: the 'Route de la Corniche' (Dakar); the 'Boulevard Lagunaire' (Abidjan, Porto-Novo); and the 'Boulevard des Iles de Loos' (Conakry). Similarly, there were names that referred to Aviation (l'Aviation); the 'Railway Station' (la Gare); the 'Trading Activity' (la Commerce), the 'Market' (le Marché), the 'Hospital' (l'Hôpital), the 'Radio Station' (la Radio), etc. Their average proportion comes to 23.8% of our total corpus, with two special cases that appear in column A, namely their under-representation in Bamako and Bobo-Dioulasso, and their over-representation in Lomé. On the whole, this mode of naming, which was also in use during the pre-colonial period, has been commonly adopted by the urban residents due to its practical significance.

The other themes, which accentuated the colonial presence, composed more than three quarters of the total names listed, i.e., 76.2%. Amongst them, references to metropolitan France in terms of general history represent an average of 23.2% (see column B). Dakar constitutes a special case with 32.1% of names attributed to these aspects, because it was the federal capital of the AOF in particular and of sub-Saharan Africa in general, and because it which was already occupied by the

French in the seventeenth century. There we note the commemoration of renowned Frenchmen or events of the kind that can only rarely be found in the neighbouring capitals. For instance, 'rue Thiers', 'avenue Gambetta', 'rue Jules Ferry', 'avenue Jean Jaurès', 'rue de Tolbiac', and also references to the French Revolution such as the streets of Fleurus and Valmy.[13] Dakar was also the only capital which toponymic inscriptions attributed names to two writers: Emile Zola and Victor Hugo. On the other hand, references related to the twentieth century are found in all the colonial cities examined.

Political personalities are a dominant category in most of the cities: General de Gaulle, who is honoured in all the territories; and also the presidents of the French Republic (for example, Doumergue in Porto-Novo, Paul Doumer in Dakar, Vincent Auriol in Bamako); and French ministers of foreign affairs (such as Albert Sarraut in both Dakar and Bamako). The two World Wars are evoked through military men (Maréchal Joffre and Maréchal Foch in Dakar, Bamako and Bobo-Dioulasso) or certain important events (rue du Dix-huit Juin in Dakar, rue du Dix-huit Juin 1940 in Bamako; avenue de Bir-Hakeim in Abidjan, Conakry and Bobo-Dioulasso; avenue de Strasbourg and Alley of Toulon in Bobo-Dioulasso; rue d'Alsace-Lorraine in Lomé, rue de l'Yser and boulevard de la Libération in Dakar, etc.). The medical domain is represented by Pasteur whose name has been given to an avenue in Dakar and a boulevard in Conakry, and Doctors Roux and Calmette each have a road in Dakar. Conakry, for its part, has dedicated two main roads to pioneers of aviation, Guynemer and Mermoz, and also a boulevard to the pioneer of the law against slavery, Victor Schoelcher. We note that at the end of the 1950s Franklin Roosevelt was the only foreign personality who had his name given to a boulevard, and that only in Dakar. France is directly cited only in Cotonou where a boulevard bears its name. In all other places, France is indirectly cited and in concrete terms, mainly through individuals or events, including the glorification of concepts referred to Liberty (boulevard de la Liberté, Dakar) and the Republic (boulevard de la République, Bamako and Dakar) (Fig. 1.4). Ironically, the latter acquired just a single avenue in Abidjan, and only after the independence of Ivory Coast.

In fact, the majority of names was related to aspects of colonial domination and dedicated to personalities or events concerning Africa. This category amounts to 53% of the corpus in total, with very minor variations in different cities, as shown in Table 7.2. We have gathered under a subdivision (column C) the names that

[13]Indeed, the reference to the Revolution in the colonial situation is somewhat unexpected, as this situation was characterised by a false *fraternité*, a denied *égalité*, and an sbsence of *liberté* – as noticed Paul Rabinow in his *French Modern* (Cambridge: MIT Press, 1989), p. 278. It seems, however, that in the early stages of the policy of assimilation in French West Africa, and especially before the First World War, there was no apparent contradiction in such notions, and Africans were expected to love France and Africa simultaneously. For this question see Alice Conklin, *A Mission to Civilize: The Republican Idea of Empire in France and West Africa, 1895–1930* (Stanford: Stanford University Press, 1997), pp. 248–249 [the Editor].

were very popular, that is, names that were found in many capital cities and urban centres because of their role in French expansion and influence in the region, or that were concerned with a particular action in only one of these cities. Amongst others, for instance, are Louis Faidherbe, Joseph Gallieni, Louis Gustave Binger and René Caillié. There are also Governor Generals of AOF, such as François Joseph Clozel, William Ponty, Ernest Roume, and Joost Van Vollenhoven. We can include in this collection the 'Tirailleurs sénégalais', for whom one avenue had been named in Bamako, as a tribute to all black African troops. However, this group is very marginal and represents only 11.6% of our corpus.

The category with the highest average (41.4%) is roads that bore the names of personalities, mostly French, who were involved in local life. Africans were a very small minority among them (4.2% in total), and were even nonexistent in certain urban areas (column E). Rarely cited, they were chosen for their role and/or for their devotion to France. For example the 'avenue du Moro Naba' in Ouagadougou, 'avenue de l'Almamy Bali Siaka' in Conakry, or even the 'avenue Blaise Diagne' and the 'rue Galandou Diouf' in Dakar.

On the other hand, names of Frenchmen dominant in this category, especially of military and administrative officials. Homage was rendered to explorers and/or to those who carried out of the conquest of the territory. To cite a few examples, the 'place Jean Bayol' in Dahomey (Benin); and the 'avenue du Général Dodds' and the 'avenue Victor Ballot' in Porto-Novo; as well as the 'boulevard du Général Dodds', the 'rue du Gouverneur Bayol' and the 'rue de l'Amiral Cuverville' in Cotonou. In Bamako, French Sudan (Mali), there are streets named after 'Borgnis-Desbordes' and 'Archinard'; and in Lomé, Togo, the streets after 'Lieutenant-Colonel Marroix' and 'Lieutenant Guillemard.' At the same time, most of the governors have their names given to streets of the main urban centres of the colony they administered. In the case of Dakar, these consisted of: François Blanchot, who governed Senegal from 1789 to 1801 and from 1802 to 1807; Auguste Léopold Prôtet (1850/1854), who had his name given to the main square in the city's colonial hub; Jean Bernard Jauréguiberry (from 1861 to 1863); Émile Pinet-Laprade from 1865 to 1869, etc. The commemoration of certain administrators was inscribed in the same manner. In Cotonou, for example, the streets after the 'Administrator-in-Chief Proche' and the 'Administrator Carr.' There are less members of other professions. Of them the doctors were best represented, such as, among others, the Doctors Jamot and Crozet in Abidjan; Guillet, Roux and Thèze in Dakar; and Quintin in Bamako. Then come businessmen, such as Verdier in Abidjan; Sanderval in Conakry; Victor Régis and Cyprien Fabre in Porto-Novo. Next, thought rarely represented, were religious personalities, such as Monseigneur Steinmetz in Cotonou, or Francis Aupiais in Porto-Novo (Fig. 7.2).

By according a key place to colonial players active in the local arena, the contemporary urban toponymy has contributed to fixing the image of French domination over the indigenous populations, a colonisation that was present in daily life.

Fig. 7.2 'Rue Francis Aupiais' in the old city centre of Porto-Novo, after a catholic missionary-ethnologist (1887–1945) active in Dahomey (now Benin), the constitutor of the Dahomey art collection for the Vatican (Photo by Claire Simonneau)

7.3 An Ambiguous Re-appropriation of the Toponymic Model Around the Year 1960

After independence, the decolonisation of toponymy affected the cities in an uneven manner. In fact, very few of these cities changed radically in terms of place naming. In the case of Ségou (Mali), for instance, the municipal authorities decided, from 21 September, 1961, that the 'boulevard Archinard', the 'avenue des Fêtes', and the streets 'Carrier', 'Hacquard', 'Desbordes', 'Combes', 'La Marne', 'Gallieni' etc. – would bear names that are more appropriate to post-independence, such as, inter alia, the 'boulevard de l'Indépendance', the 'rue El Hadj Omar', the stadium 'Modibo Keita', and 'PLACE des Martyrs du Colonialisme.'[14]

In most of the cases, the end of colonial domination was celebrated by renaming of main roads. In Abidjan, for example, the 'place Lapalud' became the 'place de la République', while the 'boulevard Antonetti' was changed to 'boulevard de la République.' In Dakar, the 'place Prôtet' was replaced with the name 'place de l'Indépendance' – the latter name is found all over the former colonies in question, in one form or another. Also extremely popular was the commemoration of African

[14]Francis Simonis, 'Une communauté française expatriée, les Européens de la region de Ségou au Mali' (PhD thesis, Paris VII University, 1993).

Fig. 7.3 Two signposts of 'Rue Parchappe' from the colonial and post-colonial periods, a testimony of the persistence of the French colonial legacy (Photo by Liora Bigon)

personalities who had played a role in the history of their country. For example, in Abidjan, avenues were dedicated to the comrades of Houphouët-Boigny, who were referred to as 'militants de la première heure' (i.e., 'early activists'), such as 'Biaka Boda' and 'Dadié Gabriel.'

Nevertheless, the capital cities and main urban areas conserved their colonial toponymy to a large extent, at least during the first decade following their independence, thus generally accepting the colonial heritage (Fig. 7.3). This tendency might be related to the French 'assimilationist' policy, which was always present even in the subsequent phase of 'association.'[15] Thereafter, the diverse political orientations of African countries also had repercussions on the urban toponymy. In Cotonou or in Conakry, for instance, the 'héros du socialisme' were highlighted. The comparison

[15]Scrutinising the doctrine of assimilation *per se* is beyond the scope of this chapter. For more on this French colonial policy, and its successor policy of association, see these 'classical' sources: Raymond F. Betts, *Assimilation and Association in French Colonial Theory, 1890–1914* (New York: Columbia University Press, 1961); Conklin, *A Mission to Civilize*; Wesley Johnson, *The Emergence of Black Politics in Senegal: The Struggle for Power in the Four Communes, 1900–1920* (Stanford: Stanford University Press, 1971); Martin D. Lewis, 'One Hundred Million Frenchmen: The Assimilation Theory in French Colonial Policy, in Robert O. Collins (ed.) *Problems in the History of Colonial Africa, 1860–1960* (Englewood Cliffs: Prentice Hall, 1970), pp. 165–78 [the Editor].

of Dakar's plan of 1983[16] with that of 1958 shows up some very minor differences, including toponyms mentioned above, 'place Prôtet', the transformation of 'rond-point de l'Étoile' to 'place Tascher' and the avenues 'Gambetta' and its extension 'Maginot' to 'Président Lamine Gueye' – were almost the only ones. Amongst the other minor changes was the modification of the names of few of Dakar's main roads. For example, the 'avenue William Ponty' was replaced for the benefit of 'André Peytavin', and the 'avenue Jean Jaurès' became the 'avenue Jean XXIII', starting from the 'Place de la République.' In total, little of the colonial toponymy disappeared.

In fact, whatever the degree of changes to the toponymy of the main arteries and boulevards – these changes corresponded with political motivation to create places of national solidarity – the essential characteristics of the colonial model were preserved. Nevertheless, the habits of the citizens did not get modified in so far and they continued to use the same methods of spatial identification as were used before independence. Moreover, this new urban toponymy has been less or rarely employed by the African populations, while the diminution of the number of Europeans whom it normally served contributed to the disuse.[17] In many cases, the newly introduced post-colonial memory became also forgotten because it was not inscribed on the signposts and name plates affixed on the corner of the streets roads: the colonial name plates were not actually replaced.

[16]Victor Franco, *Au Senegal*, series guides à, (Paris: Hachette, 1983), pp. 126–127.

[17]As noted by Francis Simonis regarding Segou: Simonis, 'Une communauté française.'

Chapter 8
A Toponymy of Segregation: The 'Neutral Zones' of Dakar, Dar es Salaam and Kinshasa

Luce Beeckmans

Abstract This chapter explores the politics of the naming of the separation zones or 'neutral zones' between Africans and Europeans in three colonial African cities: British Dar es Salaam, French Dakar and Belgian Kinshasa. By showing that in the cities discussed, colonial powers not only used place-naming as a strategy to impose their dominance, but also to legitimise and camouflage the segregationist intentions of their urban policies, we aim to add another layer to the study of the political practice of place-naming within a colonial context. In addition, this chapter uses a toponymic analysis to understand the importance of transnational connections and networks in the introduction of racial segregation in Dakar, Dar es Salaam and Kinshasa, as well as in the accompanied legitimising discourse. Finally, we will illustrate how this legitimising discourse eventually resulted in official place names. Yet, while many of the resulting names are still inscribed in the toponymic landscape of modern Dakar, Dar es Salaam and Kinshasa, others have disappeared as a result of spontaneous or symbolic renaming operations in the eras before and after decolonisation.

Keywords Racial segregation • African colonies • Colonial toponymy • Neutral zones • Urban environments

8.1 Introduction

It is strange to encounter an open space in the middle of Tanzania's congested capital, Dar es Salaam. Similarly, one might wonder why there exists a golf course and a zoo besieged by traffic jams in the bustling city centre of Kinshasa, capital of the Democratic Republic of the Congo. And while Senegal's capital city Dakar does not feature an empty space at its heart like these other cities, there is something odd about its messy triangle of streets, containing a mosque and two football stadiums. The key to understanding the intriguing anomalies in the urban geographies of these capital cities lies in tracing back the histories of their place names.

L. Beeckmans (✉)
Architecture and Urban Planning, Ghent University, Ghent, Belgium
e-mail: luce.beeckmans@ugent.be

L. Bigon (eds.), *Place Names in Africa*, DOI 10.1007/978-3-319-32485-2_8

In doing so, this chapter aims to contribute to the wave of critical reformulations occurring in place-name studies over the last decade in fields as varied as governmentality studies, urban anthropology and cultural geography. A common vantage point for many of these studies is the reading of place-naming as "a political practice par excellence of power over space."[1]

Indeed, place-naming has often been an endeavour to render an unfamiliar and chaotic toponymic and geographic landscape more orderly and manageable.[2] In the colonial context several authors have pointed to the "key role of naming in the colonial silencing of indigenous cultures", as well as to the many toponymic struggles that have resulted from this oppression, ranging from the spontaneous use of alternative names to renaming campaigns in both colonial and post-colonial eras.[3] In this chapter we would like to add another layer to the study of the political practice of place-naming. By drawing on a wide variety of primary and secondary sources we will argue that in the three African cities discussed, colonial powers not only used place-naming as a strategy to impose their dominance, but also to legitimate and camouflage the segregationist intentions of their urban policies. As such, this chapter aims to respond to scholarship that has become 'too predictable and formulaic in its repetitious invocations of toponymic domination and resistance.'[4] In addition we will illustrate how this legitimising discourse eventually resulted in official place names. Yet, while many of the resulting names are still inscribed in the toponymic landscape of modern Dakar, Dar es Salaam and Kinshasa, others have disappeared as a result of spontaneous or symbolic renaming operations in the eras before and after decolonisation.

8.2 Transnational Exchange of Place-Name Legitimisations

After the First World War, when many African cities encountered a significant influx of Europeans, colonial governments in Dakar, Dar es Salaam and Kinshasa proposed the establishment of a physical partition between Africans and Europeans.

---

[1]Jani Vuolteenaho and Lawrence D. Berg, 'Towards Critical Toponymies', in Jani Vuolteenaho and Lawrence D. Berg (eds), *Critical Toponymies: The Contested Politics of Place Naming* (Burlington: Ashgate, 2009), pp. 1–17 (p. 1).

[2]Reuben Rose-Redwood, Derek Alderman and Maoz Azaryahu, 'Geographies of Toponymic Inscription: New Directions in Critical Place-Name Studies', *Progress in Human Geography*, 34, 4 (2010), pp. 453–470 (p. 454).

[3]Vuolteenaho and Berg, 'Towards Critical Toponymies', p. 2. With regard to the colonial context, see among others: Liora Bigon, 'Names, Norms, and Forms: French and Indigenous Toponyms in Early Colonial Dakar, Senegal', *Planning Perspectives,* 23 (2009), pp. 479–501; Garth Myers, 'Naming and Placing the Other: Power and the Urban Landscape in Zanzibar', *Tijdschrift voor Economische en Sociale Geografie,* 87 (1996), pp. 237–246; Brenda Yeoh, 'Street Names in Colonial Singapore', *Geographical Review,* 82 (1992), pp. 313–322.

[4]Rose-Redwood, Alderman and Azaryahu, 'Geographies of Toponymic Inscription', p. 466.

So although a strict separation between Africans and Europeans is often assumed to have been an undeniable feature of the colonial city in sub-Saharan Africa from its early beginnings, it is clear that it was cohabitation and not segregation that was the rule before the war. Only in a very small number of cities was racial segregation imposed beforehand, while in most cities segregation along racial lines was introduced to the city layout *a posteriori* by the implementation of a building-free zone between Africans and Europeans. In this chapter we will argue that the study of the place-naming of the zones planned to separate Africans from Europeans is crucial to understanding the similarity in interventions in the urban fabric of Dakar (from 1914 onwards), Dar es Salaam (from 1924 onwards) and Kinshasa (from 1932 onwards), despite major differences in French, British and Belgian colonial ideology and urban policy.

An important motivation for colonial governments for making the shift from fairly pragmatic cohabitation to a strict separation between Africans and Europeans was a desire to materialise colonial dominance in the built space. To some extent the anticipated separation was thus disconnected from race, even though in the African context it could not but also result in racial segregation. A separation between Africans and Europeans also allowed the European minority to establish a comfortable home, a 'chez-soi' in the colony.[5] Apart from enabling better social and military control, putting a distance between colonised and coloniser was felt to offer a solution to the various inconveniences felt by Europeans living alongside Africans. These complaints were often accompanied by a moral overtone: not only was the nocturnal sound of tam-tams found to be deeply annoying, but promiscuity and racial intermingling were often thought to go hand in hand. Finally, hygiene and health also formed a major incentive to implement a physical separation between Africans and Europeans. Yet, while this preoccupation with health and hygiene was quite understandable in a continent that was often nicknamed the 'White man's Grave',[6] epidemics and unsanitary conditions were also mistakenly associated with race in colonial Africa.

Although several arguments for segregation along racial lines can thus be viewed as inherent to the colonial enterprise, racial segregation was far from easy to reconcile with colonial ideologies or policies of international bodies such as the League of Nations, predecessor to the United Nations. Although British indirect rule, with its dual administration for Africans and Europeans, compared to the French policy of assimilation, with both Europeans and an elite of assimilated Africans enjoying equal rights to some extent, provided an excellent basis to legitimise racial segregation, the British in Dar es Salaam just like the French in Dakar felt forced to

---

[5]Odile Goerg, 'De la ségrégation coloniale à la tentation sécessionniste', in Laurent Fourchard and Isaac Olawale Albert (eds), *Security, Crime and Segregation in West African Cities since the 19th Century* (Paris: Karthala, 2003), pp. 245–262 (p. 252).

[6]Odile Goerg, *Pouvoir colonial, municipalités et espaces urbains. Conakry-Freetown, des années 1880 à 1914* (Paris: L'Harmattan, 1997), p. 39.

further rationalise racial segregation with a sanitation discourse.[7] Even the Belgians, who applied a policy that was greatly inspired by both British and French rule, but that was nonetheless overtly racial, used sanitation arguments to implement racial segregation in Kinshasa. In all three colonies similar explanations underpinned the sanitation discourse, such as the statement that a physical distance would prevent malaria mosquitoes flying over from the African quarter (where native children were assumed to be particularly infected by germs) to the European quarter, or that it would prevent the contamination by germ-ridden rats "as these are less likely to move from area to area over an open space" – all explanations which were far from scientifically proven, and were even broadly contested by empirical observations.[8] This shared use of a sanitation discourse thus shows the controversial character of the planned intervention, but also suggests a certain transnational exchange with regard to the implementation of racial segregation in sub-Saharan Africa.

Toponymy is an interesting entry point to understanding the importance of transnational connections and networks to the introduction of racial segregation in Dakar, Dar es Salaam and Kinshasa, as well as to the accompanying legitimising discourse. Indeed, a remarkable similarity exists in the place-naming of the separation zones between Africans and Europeans in all three capitals: the 'Neutral Zone' in British Dar es Salaam and the French equivalent 'Zone Neutre' in French Dakar and Belgian Kinshasa. While this term was originally used in a conceptual manner without capitalisation, it gradually became a true place name, taking on the capitalisation of a proper noun. Although possibly the term itself could not be more ideologically neutral, the intentions were certainly anything but. Several archival records highlight the powerful influence of British and South African sanitation experts with regard to the 'neutral zone', as well as a significant transnational dialogue between the three colonial powers, as we will illustrate now.

After the First World War, when Tanzania (then Tanganyika) became a League of Nations mandate under British rule, the British reintroduced the German idea of implementing a 'Tote Zone' ('Dead Zone') of 256 m between Europeans and non-Europeans.[9] But the British carefully avoided any mention of the German model, as the German coloniser had often been accused of being openly racist, both by the League of Nations and others. Instead, the British referred to the theories of British and South African sanitation experts, in particular those of Professor William J. R. Simpson, one of the founders of the London School of Hygiene and Tropical

[7]For a comparison between French and British rule, see: Véronique Dimier, *Le gouvernement des colonies, regards croisés franco-britanniques* (Brussels: Editions de l'Université de Bruxelles, 2004).

[8]Quote from: Tanzania National Archives, Dar es Salaam (hereafter TNA), TNA 3088, Town Planning and Building Committee to Chief Secretary, undated (around 1923). See also: Archives Nationales du Sénégal, Dakar (hereafter ANS), ANS H55; Africa-Archives, Brussels (hereafter AA), AA, GG 7.052.

[9]Franck Raimbault, 'L'évolution de l'espace péri-urbain à Dar-es-Salaam durant la colonisation allemande (1890–1914)', in Bernard Calas (ed.), *De Dar es Salaam à Bongoland* (Paris: Karthala, 2006), pp. 265–295 (p. 103).

Medicine, who had a major professional history in British India (Calcutta, 1886–1897), as well as in South Africa (1900–1901), West Africa (1908) and East Africa (1913–1914).[10] Simpson was a convinced advocate of racial segregation as a pillar of sanitation policy, and was appointed official advisor to the Colonial Office in London in 1913. He can be considered father to the 'neutral zone' concept, although it later became appropriated by many others, for instance Lord Lugard in Nigeria, who from 1919 onwards often used the term 'neutral zone' as an alternative to 'building-free zone' ('BFZ').[11]

Simpson's visit to neighbouring Uganda did not go unnoticed by the Dar es Salaam administration, where the colonial government eagerly borrowed Simpson's concept. When the chief secretary of Tanzania warned of the controversy that might erupt when racial segregation was applied in cities such as Dar es Salaam, the land officer quickly replied: 'This is now recognised as a necessary sanitary measure', referencing Simpson's advice to the townships of Kampala and Jinja as regards the lack of segregation.[12] However, although Simpson advised a 'neutral zone' of 300 yards (274 m) between Europeans and non-Europeans, segregation between Asians and Europeans was considered impracticable and even undesirable in Dar es Salaam because of shared trading interests and high land prices in the business district. In order to somehow still comply with Simpson's advice, a 'neutral zone' was proposed between Africans on the one hand and Asians and Europeans on the other (Fig. 8.1).[13] One diligent government official pointed out this flagrant contradiction of Simpson's advice:

> I submitted to the [Central Building and Town Planning] Committee the point that the object which created the adoption of neutral zones in tropical towns, viz. a clear belt of 300 yards in width, separating the trading quarter and Asiatics from the European Residential Quarter- could not be applied in Tanganyika, secondly that the 'Neutral Zone' as at present situated simply separated the native town from the Indian Bazaar and Commercial Quarter and the object of its protection was lost.[14]

Yet the plan was still pursued. In other cities too, such as Calcutta or Stone Town (now Zanzibar), Simpson's advice was ill received, seen as overly fanatic and lacking local knowledge, with inventive adaptations as a consequence.[15]

---

[10]William John Ritchie Simpson, *The Principles of Hygiene as Applied to Tropical and Subtropical Climates* (London: John Bale & Sons, 1908). On Simpson: R. A. Baker and R. A. Bayliss, 'William John Ritchie Simpson (1855–1931): Public Health and Tropical Medicine', *Medical History*, 31 (1987), pp. 450–465; William Cunningham Bissell, *Urban Design, Chaos, and Colonial Power in Zanzibar* (Bloomington: Indiana University Press, 2011), p. 177–178; Philip D. Curtin, 'Medical Knowledge and Urban Planning in Tropical Africa', *The American Historical Review*, 90, 3 (1985), pp. 594–613 (pp. 610–612); Robert Home, *Of Planting and Planning: The Making of British Colonial Cities* (Oxford: Alexandrine Press, 1997), pp. 47, 78–81.

[11]Home, *Of Planting and Planning*, p. 130.

[12]TNA 3152, land officer to chief secretary, 18 October 1920.

[13]TNA 3152, land officer to chief secretary, 18 October 1920.

[14]TNA 3088, land officer to chief secretary, 25 March 1924.

[15]Home, *Of Planting and Planning*, pp. 78–81; Bissell, *Urban Design*, p. 177.

Fig. 8.1 Archival document showing the proposed neutral zone in Dar es Salaam in 1920 between the African quarter (*left*) and the rest of the city (Indian and European quarters on the right) (Courtesy of Tanzania National Archives)

In Kinshasa too the introduction of a 'neutral zone' between European and African quarters was found incompatible with local interests. Although the Sanitary Office proposed a 'neutral zone' of 3,000 m, a width which was later decreased to 800 m, the Municipal Council ('Comité Urbain') argued that 400 m would suffice, as this would already entail a number of expropriations. Again, inspiration was found elsewhere, namely in the proceedings of *The Sanitary Conference of Chief Health Officers* held in Cape Town from 15 – 25 November 1932 by the League of Nations, in which some Belgian officials had participated:

> I have the honour of informing the Governor that the Sanitary Conference of Chief Health Officers (organised by the League of Nations) has established in its session of November 1932 some fundamental principles of prophylaxis [prevention] of infectious diseases in Africa. With regard to yellow fever in particular its conclusions recommend: "Segregation of the residential quarters of the highly receptive European population from the native towns by a building-free zone of 400 metres."[16]

Even though the Municipal Council decided that the expropriation of Africans was relatively inexpensive and therefore a viable policy, in practice it proved very difficult because no accommodation had been prepared for the evicted people. As a result the 'neutral zone' shrank further, to a mere 250 m, again showing that practical

[16] AA, GG 7.341, Doctor Tavernari to governor, 12 June 1933 (author'–s translation from the French).

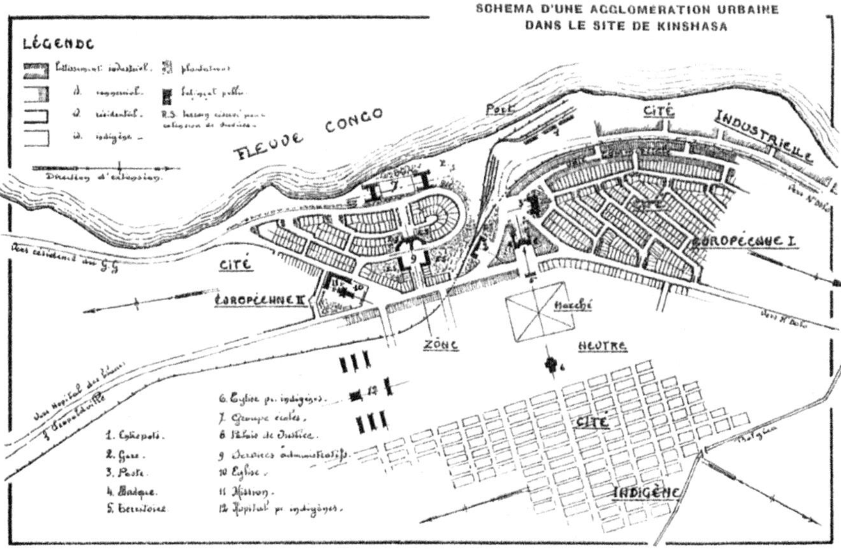

Fig. 8.2 Scheme by the architect René Schoentjes (1932) for a neutral zone in Kinshasa between the European quarter (*top*) and African quarter (*bottom*) with their distinct street layouts (Courtesy of Royal Academy for Overseas Sciences, Brussels: M. R. Schoentjes, 'Considérations générales sur l'urbanisme au Congo belge', *Institut Royal Colonial Belge, Bulletin des séances*, 4, 1 (1933), pp. 528–572)

concerns often prevailed over the pseudo-scientific knowledge that legitimised the 'neutral zones.' In 1933, almost simultaneously with such local developments, the architect René Schoentjes from the Ministry of Colonies in Brussels proposed the implementation of a 500-m neutral zone. Schoentjes' plan was never seriously taken into consideration by the Municipal Council because it required the demolition of the newly established commercial district between the African and European quarters, but it is important to note that Schoentjes also used the term 'zone neutre', and in doing so contributed to its introduction into Belgian colonial planning jargon (Fig. 8.2).[17]

Similarly in Dakar a 'neutral zone' was planned after the outbreak of bubonic plague in 1914. However, the proposed zone differed from those in Dar es Salaam and Kinshasa as it did not entail the insertion of a building-free zone (or *zone non aedificandi*) in the existing urban fabric. Instead it consisted of the forced removal of Africans to a new 'village de ségrégation' or 'village stérile' (the later Médina) at a distance of 800 m from the Plateau (Fig. 8.3). To justify its decision to introduce residential segregation between Africans and Europeans in Dakar, the Sanitary Office stated that this already formed an established practice by all other

[17]René Schoentjes, 'Considérations générales sur l'urbanisme au Congo belge', *Institut Royal Colonial Belge: Bulletin des séances*, 4, 1 (1933), pp. 528–572.

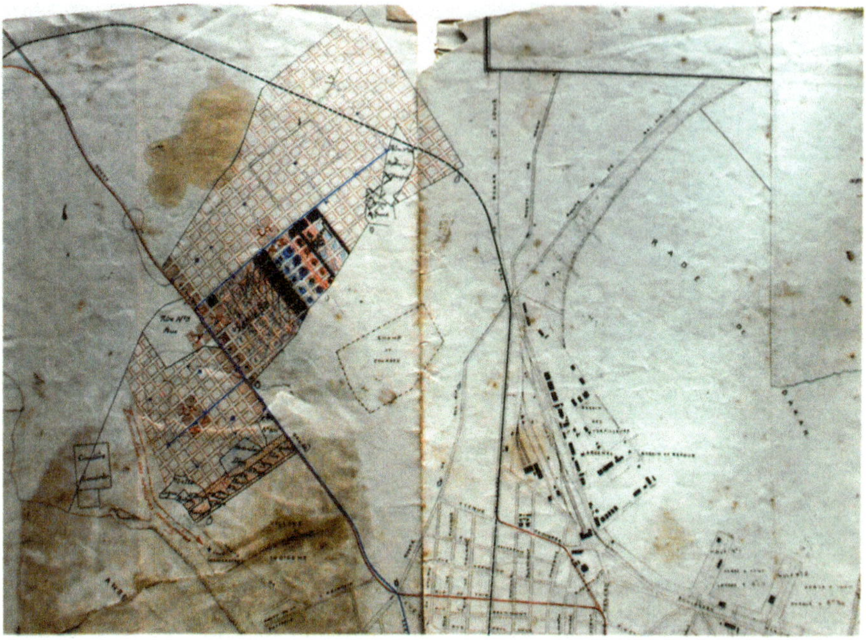

Fig. 8.3 Archival document showing the proposed Médina for Africans (*top left*) separated from the Plateau (*bottom right*) by an 800-m neutral zone in Dakar around 1914 (Courtesy of Archives Nationales du Sénégal, ANS H20)

administrations in Africa at that time, which was far from true, but presumably referred to South African examples: "This recommended measure is applied by all foreign governments and has the objective of avoiding contamination of the European quarters by the neighbouring black villages that are or could in future become reservoirs of malaria, yellow fever and plague.[18] Indeed, at the beginning of the twentieth century a sanitation discourse was already being used as a pretext and legitimation for racial segregation in South Africa, a phenomenon dubbed the 'Sanitation Syndrome' by Maynard W. Swanson and many others after him.[19] In Dakar this sanitation syndrome did not result in the creation of separate settlements for Africans miles from the city, as was the case in South Africa, but in a situation that was very analogous to those in Kinshasa and Dar es Salaam.

In Dakar, Dar es Salaam and Kinshasa the strong influence of travelling British and South African sanitation experts can be noticed after the First World War. Through influential manuals, international and above all intercontinental confer-

[18]ANS H20, 'Rapport sur la surveillance sanitaire de la ville de Dakar', 20 Decembre 1912 (author's translation from the French).

[19]Maynard W. Swanson, 'The Sanitation Syndrome: Bubonic Plague and Urban Native Policy in the Cape Colony, 1900–1909', *Journal of African History*, 3 (1977), pp. 387–410.

ences such as the *Conference of Principal Medical Officers and Senior Sanitary Officers* in Lagos in 1912, the *Inter-Colonial Conference on Yellow Fever* in Dakar in 1928 and the aforementioned *Sanitary Conference of Chief Health Officers* in Cape Town in 1932, these 'all-purpose experts' turned racial segregation, and in particular the implementation of 'neutral zones', into a legitimate sanitary measure, with considerable impact on town planning.[20] Under the influence of the discipline of Tropical Medicine, racial segregation evolved from a temporary solution in the battle against infectious diseases, as was the case in Dakar (where the 'Institut Pasteur' was established in 1923), to a permanent prevention measure (often called *prophylaxy*), as was the case in Dar es Salaam and Kinshasa, which had never even faced an outbreak of infectious disease. Therefore, even though epidemics are foremost medical phenomena, in the colonial context they clearly also functioned as political constructions and ideological instruments.[21]

This was clearly the case in Dar es Salaam and Kinshasa, where imagined diseases formed the basis for racial segregation. But even in Dakar the otherwise incomprehensible actions of the French colonial government to the outbreak of plague suggest a political response to disease. Only 10 days after the election of first African deputy Blaise Diagne on 13 May 1914, the colonial government decided to quarantine the African area to the west of the Plateau with a cordon sanitaire, even though the first cases of infection were found to the east of this zone.[22] For that matter an interesting confusion of tongues arose from this, as the 'neutral zone' of Dakar nowadays is still often referred to as the 'cordon sanitaire.'[23] While this term was indeed used by the French in North Africa to refer to separation belts at that time,[24] in archival records it exclusively refers to the boundary of the quarantined zone on the Plateau, and as a consequence did not have any width. Moreover, the question remains as to why Africans could not simply return to the Plateau once they had been declared healthy, but had to move to the African-only Médina after being exposed to several, often humiliating, sanitary measures, such as the burning of huts, the disinfection of possessions with sprays (a measure of questionable

[20]For a broad comparative perspective, see: Curtin, 'Medical Knowledge and Urban Planning in Tropical Africa.' The term 'all-purpose experts' is borrowed from Home, *Of Planting and Planning*, p. 125.

[21]Myron Echenberg, *Black Death, White Medicine: Bubonic Plague and the Politics of Public Health in Colonial Senegal, 1914–1945* (Portsmouth N.H.: Heinemann, 2002), p. 67.

See also: Elikia M'Bokolo, 'Peste et société urbaine à Dakar: l'épidémie de 1914', *Cahiers d'Études Africaines*, 85/86 (1982), pp. 13–46; Raymond F. Betts, 'The Establishment of the Medina in Dakar, Senegal, 1914', *Africa*, 1 (1971), pp. 143–152.

[22]ANS H20.

[23]Ambe J. Njoh, *Planning Power: Town Planning and Social Control in Colonial Africa* (London, New York: UCL Press, 2007); Liora Bigon, *A History of Urban Planning in Two West African Colonial Capitals* (Lewiston: Edwin Mellen Press, 2009), pp. 188–216.

[24]Gwendolyn Wright, *The Politics of Design in French Colonial Urbanism* (Chicago: The University of Chicago Press, 1991).

impact), compulsory vaccinations and a sojourn of 10 days in the newly built 'camp de ségrégation' in neighbouring Bel Air.

Consequently, what these cases suggest is that, in contrast to the widespread view, the French were no less segregationist than the British or even the Belgians.[25] In fact, it is exactly the shared use of the term 'neutral zone' that points to a certain transnational exchange of both segregationist practice and an intertwined place-name legitimisation within the African continent. Even more remarkable is that these colonial powers all used similar camouflage techniques to implement these 'neutral zones', as we will show below.

8.3 From Camouflage Techniques to Official Place Names, Colonial and Post-colonial Renaming

The segregationist ideologies underpinning the separation zones were already to some extent camouflaged by calling them 'neutral zones.' Implementing the 'neutral zones' without formulating them in racial terms was more difficult, but building regulations, which went intertwined with eviction orders and expropriations, most often without compensation, provided an appropriate solution. While in Kinshasa the building regulations were still drawn quite explicitly along racial lines,[26] in Dar es Salaam and Dakar building regulations were used precisely to camouflage the racial foundations of the policy. Indeed, building regulations allowed the 'neutral zone' to be described as an area between two zones to which different legislation applied, instead of an area between two racially defined population groups.

In Dar es Salaam the initial plan was to describe three zones defined by the three types of latrine: European, Asian and African:

> There is a universal agreement as to the wisdom and necessity of segregation except on the part of the Indian agitator, and he objects solely because objection is a political manoeuvre. But we can, I believe, ensure proper segregation in actual practice by means of Building and Town Regulations. For example, though an Asian may buy a plot in the European residential quarter, we can require him to build on it a house of a type which would not suit his methods of life in that we should prohibit the existence of the Asiatic conception of a latrine, or of an Asiatic bathing suite.[27]

---

[25]For this widespread view, see: J. L. L Comhaire, 'Urban Segregation and Racial Legislation in Africa', *American Sociological Review*, 3 (1950), pp. 392–397; Catherine Coquery-Vidrovitch, 'Residential Segregation in African Cities', in Steven Salm and Toyin Falola (eds), *Urbanization and African Cultures* (Durham NC: Carolina Academic Press, 2005), pp. 343–356; Odile Goerg, 'From Hill Station (Freetown) to Downtown Conakry (First Ward)', *Canadian Journal of African* Studies, 32: 1 (1998), pp. 1–31.

[26]Egide-Jean Devroey, *La Réglementation sur les Constructions au Congo Belge* (Bruxelles: Falk, 1941), p.136.

[27]TNA 3152, chief secretary to land officer, 22 December 1920.

Because of the strong contestations of Asians, as well as the ever-present fear of the League of Nations, the building regulations only made a distinction between three, vaguely described building types:

> So far as racial segregation is concerned it is pretty clear that in this territory we cannot adopt the principle of racial segregation as such, for that would lead us into a position where we should be in conflict with the terms of the Mandate [of the League of Nations]. Racial segregation not being practicable, a Standard to which all new buildings must confirm has been adopted and this, it is hoped, will secure the 'same advantages.'[28]

Apart from the 'neutral zone', three zones were described. In Zone I only residential buildings of European type were permitted, in Zone II residential and trading buildings were allowed, and in Zone III 'native quarters' could be erected. But even then the building regulations were found too explicit and in the final *Building Areas for Dar es Salaam Township* of 1924 every reference to the controversial 'neutral zone' was completely omitted, by describing the borders of the three zones with their street names: "The draft Notice for Gazette has been redrafted to exclude the words 'Neutral Zone.' The neutral zone is left as an undescribed area."[29]

Exactly the same camouflage technique was used in Dakar to disguise the 'neutral zone', except that there a distinction was made between building materials instead of building typologies. Although the building regulations issued after the outbreak of plague in 1914 read as an exact copy of the previous regulations of 1912, there also existed a fundamental difference. In both regulations two zones were defined: the first where only durable materials could be used and the second where significant deviations could be accepted. But if one carefully considers both regulations and, more exactly, the street names defining both zones, one can see that while in 1912 the boundary was located on the Plateau, roughly following the Rue de Vincens, in 1914 it consisted of a zone of 800 m in width.[30] This signified a crucial difference as the building regulations had evolved from a simple battle against huts on the Plateau to a straight removal of Africans to the Médina.[31] The following quote illustrates that this was certainly the objective:

> In order to achieve the pursued objective two conditions have to be realised: 1. Separation of two groups of dwellings by a building-free neutral zone – something that has been implemented today. 2. Separation of two groups of dwellers, those of Dakar that cannot live in the Médina and those of the Médina that cannot live in Dakar. This still has to be done.[32]

---

[28]TNA 3152, chief secretary to land officer, 22 December 1920.

[29]TNA 3088, land officer to chief secretary, 23 September 1924.

[30]ANS H20; ANS H22, report 'Hygiène à Dakar (1899–1920)' by Ferdinand Heckenroth on request of the Dakar mayor.

[31]Alain Sinou, *Idéologies et Pratiques d'Urbanisme dans le Sénégal Colonial* (Paris: EHESS, 1985), p. 216.

[32]ANS P190, undated note 'Village de Médina'(around 1916) (author's translation from the French).

In all three cities it was almost impossible to relocate all Africans to their designated zones. This was especially the case in Dakar, where a strong Lebou presence always prevailed around the pre-colonial village of 'Tound' ('Dunes' in the Lebou tongue Wolof, around today's Sandaga market) on the Plateau.[33] In Kinshasa too the presence of African servants in European neighbourhoods made residential segregation a quite illusory project:

> With regard to the segregation of the European and native population, it is important not to deny a 'fact' of major importance: whether or not there exists a neutral zone in Léopoldville [Kinshasa], and whether or not this neutral zone has a width of 400 metres, 800 metres or any other distance, the segregation of the European and native population only exists in theory since there is not one house in the European zone without natives living or sleeping there, day and night, and this is especially the case in the European quarters that are located many kilometres from native towns.[34]

Moreover, although colonial administrations tried to erase the name 'neutral zone' from the collective memory by camouflaging it through building regulations, it was already quite established by that time as both an unofficial and official place name, as the frequently occurring capitalisation in various reports and maps indicates. Although in the course of time the name 'neutral zone' became disconnected from its ideological underpinnings and was 'neutralised' to a certain degree in Dakar, Dar es Salaam and Kinshasa, it was also replaced by other names. In many cases this renaming occurred somewhat accidentally, with the function of the 'neutral zones' playing an important role in the renaming.

Although the 'neutral zones' as they were finally implemented differed substantially in width in Dakar (800 m), Dar es Salaam (150 m) and Kinshasa (250 m), their functions were comparable. Even though living in the 'neutral zones' was strictly prohibited, they were all endowed with a recreational character: a park, botanical garden, a zoo, a golf course, a racetrack, a football stadium or just some 'open space.' These functions often sufficed to give the 'neutral zone' a new name. In Dar es Salaam the small open strip between the African and Asian quarter was for a long time simply called 'the Open Space.'[35] Despite plans to level the Open Space, plant it with grass and trees and lay out cricket, football, hockey and ngoma grounds, after the last house was finally cleared, it remained a wasteland until long past the Second World War.[36] Yet even from the beginning of the 1930s another name was already in use, one that is still used today: 'Mnazi Mmoja', literally translated as 'One Coconut Palm.' According to some sources this name refers to an early city catastrophe in 1872, when a cyclone devastated all coconut palms, for Africans both a source and

---

[33]ANS H22, report 'Hygiène à Dakar (1899–1920)', p. 378

[34]Author's translation from the French. AA, GG 7.341, note 'zone neutre District Urbain', 7 September 1934.

[35]TNA 3088.

[36]TNA 18950 Vol. II, 'Amendments on Baker Memorandum', 1939, p.10; TNA 61/643/3 Township.

Fig. 8.4 Aerial view of Dar es Salaam around 1964 showing the building-free neutral zone (Mnazi Mmoja) in the centre, with the low-rise Swahili houses of the Kariakoo African quarter to the right, and the high-rise buildings of the Indian and European quarters to the left (Courtesy of Tanzania Information Services, Dar es Salaam)

symbol of welfare, except for one.[37] It is nevertheless very likely that a connection exists with the 'Mnazi Mmoja' zone in Zanzibar's capital Stone Town, which was implemented following Professor Simpson's visit to the island in 1914.[38]

After the war Mnazi Mmoja became a symbolic place, both for the colonial government with its legitimising development discourse, and for the anti-colonial resistance movements. Colonial welfare planning became well evidenced in the buildings inaugurated on Mnazi Mmoja: the Mnazi Mmoja Clinic, a school and the Arnautoglu Community Centre, a venue praised by the United Nations in 1952 for organising 'multiracial discussion groups' (Fig. 8.4).[39] But at the same time Mnazi Mmoja formed a battleground of contestation over place and identity. It was not

[37] Alain Ricard, *Ebrahim Hussein: Swahili Theatre and Individualism (*Mkuki na Nyota Publishers, 1995), p. 3.

[38] Bissell, *Urban Design*, p. 178.

[39] Andreas Eckert, 'Regulating the Social: Social Security, Social Welfare and the State in Late Colonial Tanzania', *Journal of African History*, 45 (2004), pp. 468–489 (p. 483); Andrew Burton

only the setting for the great strike of 1946 in Dar es Salaam, when thousands of Asians and Africans protested for better housing conditions, but also many political meetings under future president Julius Nyerere were held there before and after independence. Finally, it formed the backdrop to the independence celebrations of 1961, when it was endowed with the 'Uhuru [Freedom] Torch' and the 'Republic Fountain.' Just as in Dakar and Kinshasa, as we shall see, the former 'neutral zone' did indeed form one of the rare open spaces in the city centre where the new regime could literally materialise its nation-building aspirations, as well as break with the segregationist colonial past. However, the ambitious 1967 proposal for a new parliament building by London architects Anthony R. Davies & Partners never came to fruition.[40] In the years following, the transition of Mnazi Mmoja in an utterly symbolic place of the Tanzania regime in its quest for a clearly demarcated political and spatial identity, slowly diluted its public character, as it became completely enclosed from the 1970s onwards.

Just as in Dar es Salaam, in reality neither Dakar nor Kinshasa's building-free 'neutral zones' were devoid of buildings. Whereas in Dakar the empty 'neutral zone' did not stay vacant for long, in Kinshasa it seemed almost impossible to remove all the existing buildings. However the recreational facilities that were erected provided Kinshasa's 'neutral zone' with many new place names, such as the 'Parc De Boeck' (after the District Commissioner Ferdinand De Boeck, sometimes also called 'De Bock'), the 'Jardin Zoologique', the 'Jardin Botanique' and the vast 'Léo Golf', all names that still persist today, with the exception of the 'Léopoldville Golf' which is now the 'Golf de Kinshasa.'[41] More than anything though, the 'neutral zone' became most known as the site of the public market, which has been held there since 1943, moving to a new building, the 'Marché Central', in 1975, and which still runs there today. This market was moved from the European quarter to the 'neutral zone' where the daily presence of African traders and clients was considered unsanitary and disturbing. For reasons of safety and control the market was not placed in the African quarter, where public facilities were generally not permitted by the colonial government. The decision to place an 'unhygienic' market in the 'sanitary belt' seems contradictory, but from a pragmatic and commercial point of view it was quite commonsensical to erect the market building in the vicinity of both African and European clientele. It moreover constituted a good reason for the colonial government to crack down on informal markets along the borders of the 'neutral zone', such as the Marché de Tabora, with its name clearly referring to the many traders from East Africa in Kinshasa, as Tabora was a major trading post in

'Townsmen in the making: social engineering and citizenship in Dar es Salaam', *International Journal of African Historical Studies*, 36, 2 (2003), pp. 331–365 (p. 354).

[40] James R. Brennan and Andrew Burton, 'The Emerging Metropolis: A History of Dar es Salaam, circa 1862–2000', in James R. Brennan, Andrew Burton and Yusuf Q. Lawi (eds) *Dar Es Salaam: Histories from an Emerging African Metropolis* (Dar Es Salaam/ Nairobi: Mkuki na Nyota Publishers/ The British Institute in Eastern Africa, 2007), p. 52.

[41] Whyms (pseudoniem of Hélène Guillaume), '*Chronologie de Léopoldville de 1881 à 1956*' (unpublished manuscript, 1956).

Fig. 8.5 Kinshasa's public market around 2007 in the former neutral zone with the Park De Bock to the left, partly appropriated by traders (Courtesy of Moise Rahmani, Institut Sepharade Europeen)

Tanzania as well as the site of a memorable defeat of the Germans by the armed forces of the Belgian Congo. A decade before independence the area around the market building was already full of homes and warehouses built by Greek, Italian and Portuguese traders, who transformed it in the most lively part of the city.[42] This vivid transformation of Kinshasa's commercial district continued uninterrupted after independence, when Lebanese, Indian and Chinese traders emerged on the scene, and parts of the neighbouring Parc De Boeck became appropriated as well (Fig. 8.5). At the same time the zoo, botanical garden and golf course succeeded in holding off real estate pressures and only today seem to fulfil their legitimising sanitation objectives by operating as a lung for the congested city.

Also in Dakar the middlemen of colonial society, namely the Syro-Lebanese traders, easily found a place in the 'neutral zone' and thus saw their position quite literally spatialised. To some extent this was also the case in Dar es Salaam where the fairly heterogeneous Asian population was housed between the African and European zone, albeit not in the 'neutral zone' itself. In Dakar the 'neutral zone' had already shrank to half its 1925 size owing to lack of space on the Plateau. Beginning in the 1930s the western part of the remaining area, called Rebeuss (probably a derivation from the French 'rebeu' meaning 'immigrant from the Maghreb')

[42]Luce Beeckmans and Liora Bigon, 'The Making of the Central Markets of Dakar and Kinshasa: From Colonial Origins until Post-colonial Times', *Urban History* (published online 8 June 2015).

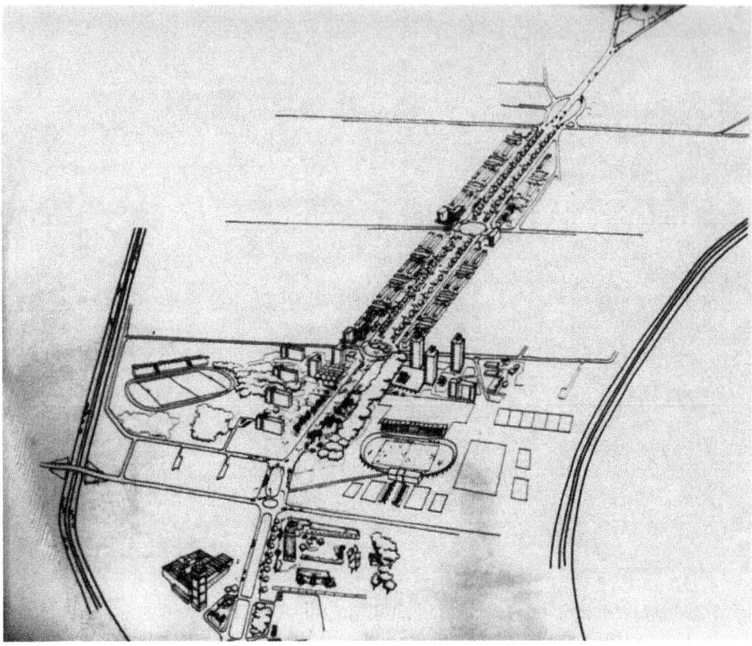

Fig. 8.6 (Non-executed) design by Michel Ecochard for the former neutral zone of Dakar, comprising several high-rise buildings, shopping centres and car parks as well as the existing mosque and football stadiums (Courtesy of Archives Nationales du Sénégal, Fonds STAGD)

was also rapidly developed.[43] To the east a series of public facilities was erected. Aside from recreational and sports facilities such as the 'Parc des Sports', the 'Stade Fédéral', and the 'Stade Iba Mar Diop', other developments were devoted to surveying (a military encampment, a police station, a prison, etc.) or fitted within a benevolent colonial discourse (the policlinic 'Institut d'Hygiène Sociale' and the 'Malick Sy' school), just as in Dar es Salaam and Kinshasa. This amalgam of facilities was joined by the 'Grande Mosquée de Dakar' in 1964, built to a design by French and Moroccan architects and inspired by the mosque of Casablanca, while plans for high-rise buildings, parking facilities and shopping centres by the French architect Michel Ecochard from 1967 were drawn up but never executed (Fig. 8.6).[44]

[43]ANS 4P 118.

[44]Luce Beeckmans, 'The Adventures of the French Architect Michel Ecochard in Post-Independence Dakar: A Transnational Development Expert Drifting between Commitment and Expediency', *The Journal of Architecture*, 19, 6 (2014), pp. 849–871.

8.4 Conclusion

To a great extent the 'neutral zones' of Dakar, Dar es Salaam and Kinshasa may be considered as the sites of unfulfilled dreams. During colonial times the intended racial segregation of the African city was never complete, as it was extremely difficult to insert a 'neutral zone' within the existing fabric of the city or keep it clear of buildings after completion. Even during the colonial era the 'neutral zones' of Dakar, Dar es Salaam and Kinshasa were, in contrast to their creators' aspirations, far from 'frozen spaces.' On the contrary, thanks to their location in-between various population groups, to various degrees these 'neutral zones' became places of physical mobility and social interaction right from their very creation.

After independence, the partly implemented or re-appropriated 'neutral zones' became symbols par excellence of colonial oppression, as well as unusual open spaces in otherwise congested city centres. As such, they became ideal sites for the new regimes to use to express their belief in socio-economic progress and modernisation through architecture, as well as proclaim the worthiness of their regime and power in the built space. In reality however, hardly any of these nation-building projects were executed. The fact that these former 'neutral zones' still mark a segregation in the urban fabric (albeit more socio-economic than racial nowadays) and that many of them still bear their original colonial names neatly illustrates the powerlessness of the new regimes in translating their conceptions of national identity to the built space. Therefore, when it comes to the former 'neutral zones' of Dakar, Kinshasa and Dar es Salaam, it can hardly be said that de-colonisation in general caused any passionate or nationalistic 'toponymic warfare', to borrow a Naftali Kadmon term.[45] Passive persistence, rather than symbolic resistance, seems to prevail with regard to the place names of these important colonial *lieux de mémoires*.

[45]Naftali Kadmon, 'Toponymy and Geopolitics: The Political Use – and Misuse – of Geographical Names', *The Cartographic Journal*, 41, 2 (2004), pp. 85–87 (p.85).

Chapter 9
Generic Terminology in Colonial Urban Contexts: Garden Cities Between Dakar and Tel Aviv

Liora Bigon and Yossi Katz

Abstract This chapter examines the various planning practices that stand behind the generic term 'garden cities' through an analysis of the inclusiveness and flexibility that are rooted in the term. While garden cities rhetoric and modes of planning in the western world are well covered in research literature, their counterparts in the southern hemisphere or in colonial contexts have gained relatively little attention. In scholarly works, in which comparative studies are rare, garden cities notions and practices have been considered simplistic at best, mere distortions of the original British models. The chapter traces and expands on the dissemination of garden city ideas in the early twentieth century from Britain to French Senegal and Ottoman Palestine. By bringing together Dakar and Tel Aviv (Ahuzat Bayit) our aim is not only to contribute to garden city historiography by an in-depth consideration of 'other' geographies. The aim is also to acknowledge the inherent dynamism that is rooted in both garden city terminology and implementation, which corresponds to a rich variety of vernacular contexts (including the global South-East). Semantically and practically, these contexts therefore constitute an integral and essential part of the global history of the garden city planning phenomenon.

Keywords Garden cities • Generic toponymy • Dakar • French Senegal • Tel Aviv (Ahuzat Bayit) • Ottoman Palestine • Colonial urban contexts

L. Bigon
The Institute of Western Cultures, The Hebrew University of Jerusalem, Jerusalem, Israel

Holon Institute of Technology, Holon, Israel
e-mail: liorin@hotmail.com

Y. Katz (✉)
Department of Geography and the Environment, Bar Ilan University, Ramat Gan, Israel
e-mail: katzyo1@013.net

L. Bigon (eds.), *Place Names in Africa*, DOI 10.1007/978-3-319-32485-2_9

9.1 Introduction and State of Study

The term 'garden city' represents one of the most influential ideas in twentieth-century urban planning. Still, there are only few comparative studies that directly deal with the question of local, regional and national variations in planning under the rubric 'garden cities.' Within these studies, comparative examination of the meaning of the term 'garden cities' in theory and practice outside of western Europe and North America in general, and of the global South-East in particular, is rare. The aim of this chapter is to expand on the changing meaning of the term 'garden city' beyond Ebenezer Howard's original ideas and their application in Britain and the northern hemisphere. This is achieved through an in-depth examination of the transnational dissemination of urban planning theories in colonial contexts, and of the local contexts of their diffusion and implementation.

The garden city model, as designed by Howard, will not be perceived here – as in the more traditional scholarly literature – as the 'original', 'normal' and 'desirable' model versus its realisations in the colonial world as apparently 'counterfeit', 'distorted' or 'simplistic.' Our analysis of a series of primary and secondary sources, as well as work in situ in two case studies in the southern hemisphere, shows quite the opposite.[1] That is, the answer to the question of how the garden city model diffused into, and was implemented in, Dakar (Senegal) under the French regime and Tel Aviv (originally 'Ahuzat Bayit') under the late Ottoman regime – exemplifies that both in practice and concept, these developments were integral and inseparable from the global developments of this model.

In early twentieth century French Dakar, the term *cité-jardin* had been utilised, semantically and practically, to further the interests of the colonial administration. As we shall see, the colonial administration used this model to meet its own needs, aiming at the creation of a prestigious image for the quarters designated for its employees, while ignoring the African urban majority. These quarters were made of wide, tree-lined streets, with low residential density. In Ottoman Palestine, Howard's ideas were accepted soon after their original publication (1898, 1902) and much before the British Mandate (1922). These ideas were embraced by Jewish immigrants from Europe, and strongly shaped the Zionist model of urban settlement. The garden city rhetoric was applied within the suburban context, and through the incorporation of certain elements which then seemed essential to the development of the national and cultural life on the basis of Zionist aspirations. In early twentieth century Palestine, similarly to the urban reality in French Dakar and in many other colonial cities in Africa and beyond, garden city ideas were used as an apparatus for residential segregation between the two main sectors/nationals, i.e., the Jews (in the new neighbourhood 'Ahuzat Bayit') and the Arabs (in nearby Jaffa).

---

[1]This contribution is based on three chapters that were written by these authors in: Liora Bigon and Yossi Katz (eds),*Garden Cities and Colonial Planning: Transnationality and Urban Ideas in Africa and Palestine*, 'Studies in Imperialism' series, eds John MacKenzie and Andrew Thompson (Manchester, New York: Manchester University Press, 2014). The chapters were merged together (by permission of MUP) and reoriented to meet the toponymic question. We have also used original visual evidence concerning both case studies that has yet to be published in the research literature.

Dakar and Tel Aviv have been selected as case studies in order to trace the processes through which the 'garden city' terminology was disseminated into various colonial situations. Such processes were often surprising and included unpredictable 'flows.' Against the background of the rich garden city historiography in the West, the contribution of this comparative chapter is twofold: first, in its preoccupation with non-western and 'even' non-anglophone contexts of garden cities as a planning phenomenon; and in the re-assessment of the meaning of the term 'garden city' considering these geographic and linguistic contexts. This is through a broadening of the more prevalent understanding of this generic term in order to arrive at a more nuanced and varied understanding. While the 'garden city' agenda had been specifically conceived as a spatial reaction in a certain place and time, it originally constituted a specific term. At the same time, because of the relative speed of the considerable geographic expansion of this agenda, and its many variations of application, this term is also generic. The insights provided by this chapter are aimed at revealing some of the meanings that are contained within the wide umbrella of garden-city terminology, and their legitimisation within the mainstream of garden city historiography.

A considerable number of books and book chapters on the history of the garden city movement and accompanied nomenclature, including some biographies on Howard himself, were focused, quite naturally, on the British experience.[2] In England, during several decades after the Second World War, there was a strong publication impetus concerning subjects related to urban management as it evolved out of pre-war garden-city developments.[3] Afterwards, relevant literature, especially

[2]For a partial list of high quality publications in English, in chronological order, see:

Walter Creese, *The Search for the Environment: The Garden City Before and After* (London: Johns Hopkins University Press, 1992 (1966)); Stanley Buder, *Visionaries and Planners: The Garden City Movement and the Modern Community* (New York and Oxford: Oxford University Press, 1990); Peter Hall and Colin Ward, *Sociable Cities: The Legacy of Ebenezer Howard* (Chichester: John Wiley, 1998); Standish Meacham, *Regaining Paradise: Englishness and the Early Garden City Movement* (New Haven, London: Yale University Press, 1999); Kermit C. Parsons and David Schuyler (eds), *From Garden City to Green City: The Legacy of Ebenezer Howard* (Baltimore, London: The John Hopkins University Press, 2002). For biographies or other works related to Howard see: Robert Beevers, *The Garden City Utopia: A Critical Biography of Ebenezer Howard* (London: Macmillan, 1988); Gordon E. Cherry, *Town Planning in Britain since 1900* (Oxford: Blackwell Publishers, 1996); Kenneth Kolson, *Big Plans: The Allure and Folly of Urban Design* (Baltimore, London, 2001); Peter Hall, *Cities of Tomorrow: An Intellectual History of Urban Planning and Design in the Twentieth Century* (Oxford: Blackwell Publishers, 1996 (1988)).

[3]'New towns', for example, and related terms: Frederick J. Osborn, *Green-Belt Cities: The British Contribution*, London (Faber and Faber, 1946); Ministry of Housing and Local Government, *The Green Belts* (London: Her Majesty's Stationery Office, 1962); Hazel Evans (ed.), *New Towns: The British Experience* (London: Charles Knight, 1972); College of Estate Management, *The Future of the Green Belt* (Occasional Papers in Estate Management, no. 5, 1974); Martin J. Elson, *Green Belts: Conflict Mediation in the Urban Fringe* (London: Heinemann, 1986); John Herington, *Beyond Green Belts: Managing Urban Growth in the 21th Century* (London: Jessica Kingsley, 1990).

in English, dealt with urban sustainability and community urban design, amongst other subjects.[4] But the main disadvantage of this literature is that it expands very little on themes beyond the global North-East. Moreover, the 'West' is also limited to North America, Germany, France, Australia and Russia – all in that order.

Against this background, the edited volume of Stephen Ward, which surveys and theorises garden city history, still constitutes one of the most important contributions in this area. Of about thirty years, it seems that this volume is a pioneer in analysis of the international arena, and though it refers to the global North, it deals with the planning and the conceptualisation of garden cities in England, France, Nazi Germany, Australia, America and Japan. It also traces the garden-city planning tradition from economic, institutional, cultural and aesthetic perspectives.[5] Inspired by Ward's book, this chapter's geographic emphasis has been directed towards the southern hemisphere and early-twentieth century colonial situations – these are concerned with Asia (Ottoman Palestine) and Africa (French Senegal).[6] The geographies in question therefore counterbalance the mainstream historiography, mainly from the well researched history of metropolitan Europe to the less researched European history beyond Europe. Within this equation, treating sub-Saharan Africa is unique.

[4]See, for instance: Hall and Ward, *Sociable Cities*; Parsons and Schuyler (eds), *From Garden City to Green City*.

[5]Stephen V. Ward (ed.), *The Garden City: Past, Present and Future* (London: Spon, 1992).

[6]Scholarly attention to the 'garden city' phenomenon in Ottoman Palestine is considerable in comparison to its counterpart in Senegal – the latter in fact hardly exists. See especially: Yossi Katz, 'The Extension of Ebenezer Howard's Ideas on Urbanization Outside of the British Isles: the Example of Palestine', *Geo-Journal*, 34 (1994), pp. 467–473; Yossi Katz, 'Garden City in Theory and Practice: the Example of Palestine in the Final Stages of Ottoman Rule', in Zeev Safrai, Yvonne Friedman and Joshua Schwartz (eds), *Studies on the Land: Studies in the History of the Land of Israel Dedicated to Prof. Yehuda Feliks* (Ramat Gan: Bar-Ilan University Press, 1997), pp. 343–353; Yossi Katz, 'Ideology and Urban Development: Zionism and the Origin of Tel Aviv', *Journal of Historical Geography,* 12 (1986), pp. 402–429; Liora Bigon and Yossi Katz (eds), *Garden Cities and Colonial Planning*; Liora Bigon, *A History of Urban Planning in Two West African Colonial Capitals: Residential Segregation in British Lagos and French Dakar (1850–1930)* (Lewiston, NY: The Edwin Mellen Press, 2009), esp. pp. 243–266. See also: Ines Sonder, *Gartenstädte für Erez Israel: Zionistische Stadtplanungsvisionen von Theodor Herzl bis Richard Kauffmann* (Hildesheim, Zürich, New York: Georg Olms Verlag, 2005); Miki Zaidman, 'Garden City – the Land-of-Israel Version: A Survey of Hebrew Garden Cities and Quarters, 1900-1948' (Ph.D thesis, the Hebrew University of Jerusalem, 2010) [in Hebrew].

9.2 The Diffusion of the 'Garden City' Rhetoric from Britain to France and on to Colonial Africa: the Case of Dakar

Unlike other early nineteenth-century planners of ideal societies, sometimes referred to as 'utopian societies',[7] Ebenezer Howard proposed a model that did not remain a theoretical vision. The renowned work (1898, 1902) of this British administrative officer, not a philosopher, became a practical contribution to contemporary urban planning.[8] This is aside from a rich secondary literature about his garden city concept and its immediate global dissemination, which often meant a loose application within other forms of urban planning and design. In short, Howard's vision and famous sketches presented the garden city as a usable alternative to the contemporary existing industrial, congested and polluted urban centres. These were subjected to massive immigration from rural areas, poverty, poor sanitation, slums and crime. Each garden city, which was also called by Howard a 'social city'[9] in the spirit of communal socialism, was designated to serve public interests by being planned in rural area and on communal land, limited in space and population. Howard's borrowing of various elements from previous thinkers in order to consolidate a unified, simple and practical urban system, led to considerable success and popularity.

As actual experiments, his ideas, in one version or another, spread quickly through Britain. Almost as quickly they influenced other European countries and the US as well. After the translation of his book into French at the beginning of the twentieth century, and the visits of French reformers to the first experiments in England, Howard's model became well known in France. The Association Française des Cités-Jardins was founded in 1904, led by Georges Benoît-Lévy, eventually its secretary and president, who fostered the garden city discourse in France in numerous publications. Other French reformers became internationally active, such as Henri Sellier, who joined the International Garden Cities and Town Planning Federation.[10] But, with the diffusion into France, the meaning of the term 'garden cities' (now *cités-jardins*) had changed in conformity with the French metropolitan urban tradition. Moreover, another layer of change had been introduced into its terminological and practical meaning with the further diffusion from France itself into the French colonial context overseas.

[7]For example Robert Owen, Charles Fourier and Saint-Simon. See Martin Buber, *Paths in Utopia* (Boston: Beacon Press, 1960), p. 6; Peter Batchelor, 'The Origin of the Garden City Concept of Urban Form', *Journal of the Society of Architectural Historians*, 28 (1969), pp. 184–200.

[8]Ebenezer Howard, *Garden Cities of To-Morrow*, edited with a preface by Frederic J. Osborn, introductory essay by Lewis Mumford (London: Faber and Faber, 1970 (1902, 1946)). Originally titled *To-Morrow: A Peaceful Path to Real Reform* (1898).

[9]Howard, *Garden Cities*, p. 138.

[10]A. Sutcliffe, 'Le contexte urbanistique de l'oeuvre de Sellier: la transcription du modèle anglais de la cité-jardin', in Katherine Burlen (ed.), *La Banlieue-oasis, Henri Sellier et les cités-jardins (1900–1940)* (Saint-Denis: Presses Universitaires de Vincennes, 1987), pp. 67–79.

While in nineteenth-century France, industrialisation and urban revolution were far less prominent than in Britain, its urban problems were largely underestimated. Georges-Eugène Haussmann was preoccupied with improving street communications at the expense of housing, and it was only after the First World War that town planning became a national concern. During the interwar period in the Paris region, *cité-jardin* rethoric had been used for suburban developments versus the previous anti-model of speculative, voluntary emerging *lotissements* (parcelling). But the purpose of these national efforts to rationalise suburban growth was not really the creation of new (garden) towns, but rather the decongestion of Paris.[11] In *cité-jardin* experiments in suburban areas there was a clear preference for apartment buildings – as opposed to the cottage as a model housing unit in Britain.[12] Similarly to the actual situation in Britain, in spite of the long socialist urban tradition in France, there was hardly any reference in France to the radical solutions offered by Howard concerning communal land ownership or the *cité-jardin* as an independent variable. This fact contributed to the application of Howard's garden city model as one form or another of *banlieues-jardins* (garden suburbs), which intensified class differences both in Britain and France.

Similarly to the early garden city developments in Britain, garden city terminology in the French planning system was used as an apparatus to legitimise urban zoning according to function. In other words, this terminology enabled the recruitment of the residential issue for the preservation of class segregation in France to camouflage urban, sanitary or social amelioration.[13] In the colonial circumstances, the socio-cultural paternalism in the mother country became racial paternalism overseas. Residential segregation on a class basis was prevalent within white society in West Africa, primarily supported by the colonial administration. This administration simultaneously established residential segregation on a racial basis between coloniser and the colonised sectors. Below we shall exemplify the particular role of using garden city rhetoric within the French colonial urban planning system in order to achieve racial residential segregation.

The strategic position of Dakar, the westernmost point in West Africa and thus a port of call on the way from Europe to South America or South Africa, was acknowledged by the French following the Crimean War and the later 'scramble' for Africa. The first town plan for Dakar was drawn up in 1862, hoping to turn the city-to-be into the capital of the French colonial empire in West Africa. The orthogonal lines of the master plan were drawn straight over the indigenous Lébou villages – these were gradually transferred to the margins of the city. The city centre

---

[11]Paul Rabinow, *French Modern: Norms and Forms of the Social Environment* (Cambridge, Mass., London: MIT Press, 1989); Annie Fourcaut, 'Débats et réalisations de l'entre-deaux-guerres ou le lotissement comme anti-modèle', in Danièle Voldman (ed.), *Les Origines des villes nouvelles de la région parisienne, 1919–1969*, Cahiers de l'Institut d'histoire du temps présent, 17 (1990), pp. 11–22.

[12]Jean Pierre Gaudin, 'The French Garden City', in Stephen V. Ward (ed.), *The Garden City: Past, Present and Future* (London: Spon, 1992), pp. 52–68.

[13]Gaudin, 'The French', pp. 55, 57, 61, 63.

became known as the 'Plateau',[14] and its regional importance increased following the designation of Dakar as the capital of the AOF (*Afrique Occidentale Français*) in 1902 (Fig. 9.1).[15] As the nerve centre of the AOF, the main function of the layout and design of the Plateau was the westernisation of the city through the adaption of French colonial urban norms and forms. Dakar was intended to fulfil an international role rather than a local one, West African rather than Senegalese. The white residential area of the Plateau constituted a model space within the larger space of Dakar as a whole, into which some Parisian elements were introduced,

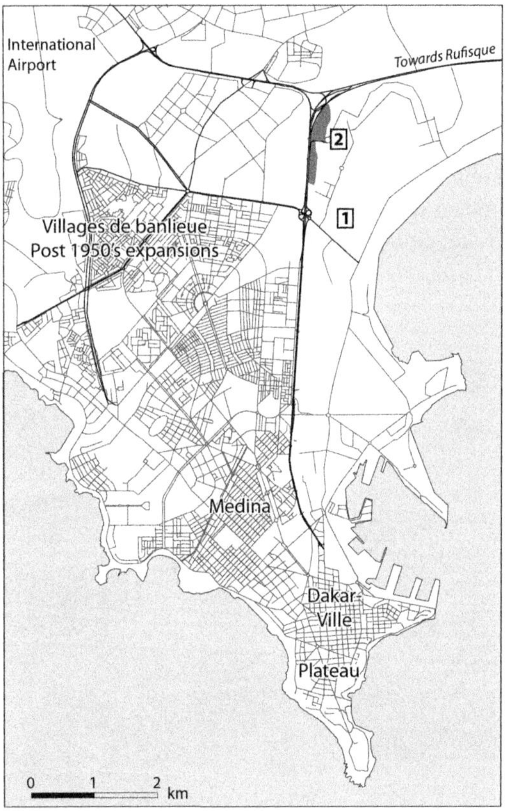

Fig. 9.1 Dakar's quarters by the 1950s. (1) Forestry Park of Hann (Hann Botanical Gardens); (2) Hann Garden City (Cité-jardin de Hann) (original drawing by Tamar Soffer)

[14]For the generic toponymy of the Plateau in the French colonies see: Liora Bigon and Ambe Njoh, 'The Toponymic-Inscription Problematic in Urban Sub-Saharan Africa: From Colonial to Postcolonial Times', *Journal of Asian and African Studies*, 50, 1 (2014), pp. 25–40.

[15]The AOF, i.e., the federation of French West Africa, was created in 1895, alongside the federation of French Equatorial Africa (AEF), to facilitate the centralist decision-making process in Paris. The AOF's overall territory amounted to 4,633,985 km^2, and included eight colonies: Senegal, French Sudan (today's Mali), French Guinea, Ivory Coast, Dahomey (Benin), Upper Volta (Burkina Faso), Niger and Mauritania.

such as Beaux-Arts aristocratic *ronds-points* and gardens. It was meant not only to impress the Africans, who, according to Eurocentric views, would learn to imitate European modes of habitation, but also to 'pacify' critics at home, especially those who denounced the colonial project.[16]

While these urban amenities were directed towards the privileged white quarters, the African residents of Dakar hardly benefited from them at all. This system of 'dual cities' was promoted by the French in North and sub-Saharan Africa, though French officials never imposed overt racial restrictions in residential matters. This is because French West Africa never experienced a permanent or massive presence of white settlers, mainly because of the tropical climate. Segregationist moves supported by Dakar's authorities were therefore relatively soft. Indeed, almost every *cité-jardin* reference in colonial times relates to the Plateau quarter. By the 1930s the 'garden city' rhetoric was extremely common in French-speaking cities in sub-Saharan Africa. The term *cité-jardin* was normally used by the French colonial administration with reference to its expatriate quarters, but it is clear, however, that more than any resemblance to Howardian notions, the *cité-jardin* in this context meant tree-lined streets, villas with verandahs and gardens, and an abundance of vegetation in general, predominantly for decoration (Fig. 9.2).

Fig. 9.2 Villa from the 1930s on Dakar's Plateau (Photo by Liora Bigon)

[16]For the 'pacification' of anti-colonial metropolitan critique see: Alain Sinou, *Comptoirs et villes coloniales du Sénégal: Saint-Louis, Gorée, Dakar* (Paris: Karthala, ORSTOM, 1993), p. 300.

Thus, wherever possible, the French ideal for housing expatriates in sub-Saharan colonies was a private residential unit with a garden, facing a straight street lined with streets.[17] The spacious, green appearance of these quarters has ben very much preserved to this day. Contemporary colonial urban planners also pointed to these physical characteristics, as clear from the reports which were published in 1932 by the International Conference of Urbanism in the Colonies and Tropical Countries, following the 1931 Colonial Exposition in Paris.[18] From Algeria to Madagascar and from Senegal to Congo Brazzaville, throughout the French dependencies, the expatriate residential form in these reports was most often labelled *cité-jardin*. The white residential area in Antananarivo (Madagascar) was described there as a "satellite garden-suburb"[19]; in Thiès (east of Dakar) it was described as "a real garden city"[20]; in Elisabethville (Belgian Congo) as "a large garden city whose greenery creates its charm"[21]; and even in Beira (Portuguese Mozambique) the white area was characterised as a desirable "garden city."[22] As stated above, one should not look here for a direct application of the British garden cities movement, but rather understand the appellation as a successful inclusion of vegetation within the white residential quarters. This can be deduced from the greenness of the metaphor. However, the indigenous green neighbourhoods or localities were rarely referred to or conceived of by Africans and Europeans alike as *cités-jardins* – a notion that is essentially occidental and expressive of essentially occidental imagery and pretensions.

Archival evidence that refers to Dakar as a *cité-jardin* supports this argument: we find the notion mentioned in the context of planting the park of Hann, or the maintenance of the villa gardens of the colonial administration on the Plateau, both in the early 1920s.[23] There is another reference, from the late 1950s, to the establishment of a new residential neighbourhood by the administration to house its

[17]Catherine Coquery-Vidrovitch, 'À propos de la cité-jardin dans les colonies', in Catherine Coquery-Vidrovitch and Odile Goerg (eds), *La Ville Européenne outre Mers* (Paris: l'Harmattan, 1996), pp. 105–126; Sinou, *Comptoirs*, pp. 307–308.

[18]Jean Royer (ed.), *L'Urbanisme aux colonies et dans les pays tropicaux*, vol. 1 (La Charité-sur-Loire: Delayance 1932).

[19]E. Weithas, 'Rapport général sur l'urbanisme en Afrique tropicale', in Royer (ed.), *L'Urbanisme aux colonies*, vol. 1, pp. 111–14 (p. 112).

[20]Weithas, 'Rapport', p. 113.

[21]René Schoentjes, 'Considérations générales sur l'urbanisme au Congo belge', in Royer (ed.), *L'Urbanisme aux colonies*, vol. 1, pp. 170–88 (p. 178).

[22]R. de Andrade, 'L'Urbanisation de Beira en Afrique orientale portugaise', in Royer (ed.), *L'Urbanisme aux colonies*, vol. 1, pp. 141–5 (pp. 144, 145). In this report the *cité-jardin* expressed colonial dreams as it had not yet been materialised.

[23]For instance: Archives Nationales du Sénégal (hereafter ANS), 4P 169, Urbanisme au Senegal: services des parks et jardins, 1923; 4P 1461, Jardins publics de Dakar et Gorée: construction, 1921–22.

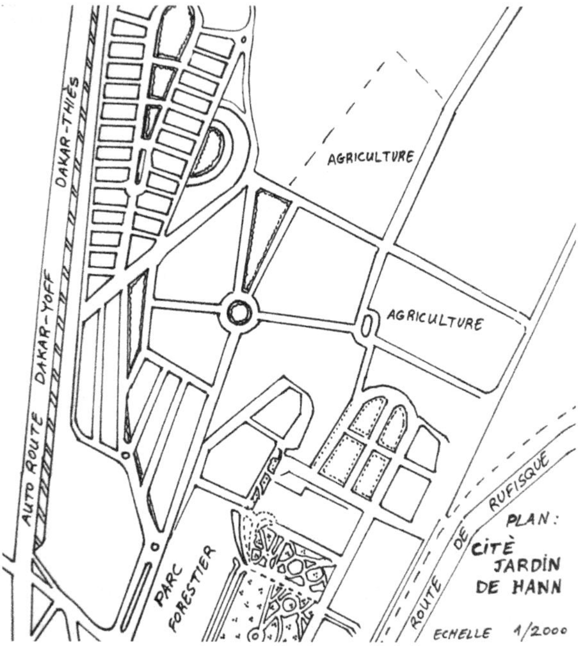

Fig. 9.3 Plan of the quarter 'Cité-jardin de Hann', 1957 (Redrawn by Liora Bigon after 4P 75, Archives Nationales du Sénégal)

employees. A part of the post-Second World War impetus of urban development in the AOF, this neighbourhood was placed north of Dakar-ville, in the triangle created by the main road to Thiès and the Forestry Park of Hann (now Hann Botanical Gardens).[24] The administrative villas were built on garden plots, arranged around squares and other areas reserved for small 'parks', and bore the title 'Hann Garden City' (Cité-jardin de Hann) (Figs. 9.1 and 9.3). This title was also derived from the association with, and physical proximity to, Hann Forestry Park and other large plots nearby that were allocated to agriculture. Hann was the name of one of the Lébou villages which was placed there on the eve of the colonial occupation of the peninsula and, following the realisation of the plan of the Plateau, was transferred farther inland. There was also an important source of ground water to be utilised for colonial Dakar, and, already by the early 1870s, an agronomic station was created there designated, inter alia, to supply trees for the main roadways of Dakar (Figs. 9.1 and 9.4).[25]

[24]ANS, 4P 75, Plan d'aménagement de la cité jardin de Hann: pièces écrites et plans, 1957.

[25]ANS, 4P 1461, Jardins publics de Dakar et Gorée: construction. Inside: Jardin des compagnies disciplinaires à Hann, 1870 (Génie, direction du Sénégal, petit atlas des batiments militaires); P167, Urbanisme de Dakar, rues et places, etc. 1901–18. Note par l'inspection de l'agriculture sur la plantation des avenues de Dakar, 28 Septembre 1907.

Fig. 9.4 French postcard from the 1910s shows the gardens of the Lébou village of Hann (Bigon's collection)

The eventual materialisation of the ideal of green space within Dakar's Plateau resulted in the resemblance of this quarter to an urban park dotted with regularly arranged villas. Like the latter Hann Garden City, Dakar's Plateau was not a garden city per se, in the sense of being an autonomous unit, nor was it really a garden suburb. While the Plateau operated as an administrative, commercial and strategic centre, it could hardly be regarded as an industrial core. This is because the colonial situation in Africa did not include any form of industrial revolution. Apart from a few white settlers' colonies, most of the colonies were reservoirs of raw materials and then markets for manufactured goods. Because industrialisation was necessarily reserved for Europe, expressions of the 'garden city' in Africa were by no means a reaction to highly industrialised urban environments. In addition, the Plateau's land nationalisation or municipalisation was part of the colonial situation and thus far removed from any socialist dimension. In fact socialist aspects contradicted the colonial situation, which was inherently anti-social. In France (and Britain) plans of *cités-jardins* were intended to upgrade the living conditions of the working class or at least were perceived as such. Yet, *cités-jardins* were never planned for the Dakarois subjects – the only sector of the colonial city that could be considered working class. Even the Office des Habitations Economiques, established by Governor General Jules Carde in 1926 in order to offer monetary assistance to Dakarois wishing to build permanent houses, completed only 20 residential units by 1945. These were the sole initiatives for tens of thousands of

Africans, in contrast to the hundreds of villas erected for French officials, and largely supported by the administration.[26]

9.3 'Garden City' Rhetoric and Urban Development in the Late Ottoman Era: the Case of Tel Aviv (Ahuzat Bayit)

We shall now turn to examine the reception of Howard's widely disseminated ideas in early twentieth-century Palestine and their influence on Ottoman era urban developments (up to 1917), with a focus on Tel Aviv. Aside from comparative remarks to some key features in the case of Dakar, it should be pointed out that two directions of 'flows' are salient in the case of Tel Aviv: first, a flow of ideas from the national to the global level. That is, the general adaptation of the British garden city models by the Zionist movement exemplifies this movement's openness to innovative and universal directions, even if they seem far from Hebraic histories and Jewish European-cum-Oriental traditions. In this sense, Zionism, which had been initially shaped in Europe, brought the garden city ideas into an old-new land in a manner that could not be realised in Europe. This movement not only absorbed urban innovations, but also considered itself progressive by pursuing a universal message of improving the world. The second direction of 'flow' moves from the global level into the national level: the utilisation of a globally spread model by its diversion to conform to specific national needs and a distinctive local context. The process of creating new Jewish national identity in Palestine needed a continuous recruitment of innovative models, symbols and ideas. The terminological and practical use in the garden city concept reflected the will to establish a model environment in both physical and moral terms.

The application of the garden city model in Palestine, normally in the form of suburban quarters and not independent cities, had much in common with Howard's notions of establishing a strong sense of community, the positioning of public buildings in the heart of the settlement and relatively small-sized residential plots. These aspects are different from the anti-social colonial situation that characterized, for instance, French Dakar. However, garden city developments amongst the Jewish sector could be also perceived as a sub-colonial situation within a wider official colonialism (Ottoman and the later Mandatory regime). This is through directing urban amenities for the benefit of one sector, while ignoring the indigenous Arab sector. There is also a colonialist ambience in the importation of the garden city terminology and imagery into the Middle Eastern environment and traditional settlement forms. These sub-colonial developments are not 'colonial' in the strict sense of an official imperial conquest of an overseas territory for the economic

[26]For more on the Office see Sophie Dulucq and Odile Goerg, *Les Investissements publics urbains en Afrique de l'ouest, 1930–1985* (Paris: l'Harmattan, 1989), p. 119. For the special conditions given by the administration to French contractors see ANS, 4P 272, Plateau, villas, 1922–23.

benefit of the metropolitan country, through, inter alia, compulsory work and military recruitment of the colonised. But it is still possible to conceive some aspects in the Zionist movement's thought and practice as colonialist.

Prior to 1917, Palestine had been under Turkish rule. More liberal than the Orthodox population of Jerusalem, the Jewish immigrants to Palestine preferred for the most part to settle in towns like the port of Jaffa rather than in agricultural settlements. This preference, particularly for Jaffa, was the result of the town's location as a port city, as well as the immigrants' natural desire to remain in the port of their arrival. Jaffa, an old Middle Eastern city and not a colonial creation such as Dakar – was also situated at the centre of the Jewish agricultural settlements founded during the late nineteenth century. But housing conditions and sanitation in early twentieth-century Jaffa were hardly adequate and considered difficult by the immigrants from established towns in Europe. Moreover, because of the large number of Jewish immigrants reaching Jaffa, it was hard to build apartments or even rent them.[27]

These conditions spurred the Jewish population there to organise and establish a modern suburb near Jaffa and to choose as their model the garden suburb, the most up-to-date European concept in contemporary town planning.[28] They had learned of Howard's model and the English and German literature written about it while still in Europe, and through Zionist leaders and other individuals in Germany who sent the literature to their representatives in Palestine.[29] However, those Jaffan Jews – an informal colonising power operating under the umbrella of the formal Ottoman regime – banded together to build a new suburb. They wished not only to accommodate material needs but also to promote Zionist aims by participating in the development of rural and urban settlement in Palestine. Thus, they emphasised the cultural and social aims of the suburb as well as its physical character.[30] According to the plan, the suburb was to be parcellated into lots measuring at least 600 square metre on which fully detached houses were to be built. The houses were to cover not more than 30 % of the site; the remainder serving for gardens and trees. Areas were to be earmarked as public gardens; roads and pavements were to be designated and paved. The houses were to have sewers and a water pipe system as well as pipes for watering gardens. Drinking water was to be drawn from wells rather

[27]David Smilanski, *With My Fellow Countrymen and Townsmen* (Tel Aviv: Yedidim, 1958), pp. 482–4 [in Hebrew]; Yossi Katz, 'Ideology and Urban Development: Zionism and the Origins of Tel Aviv, 1906–1914', *Journal of Historical Geography*, 12, 4 (1986), pp. 402–424.

[28]Central Zionist Archives (hereafter CZA), L51/52, Letter from the Ahuzat Bayit Association to the Keren Kayemet (The Jewish National Fund), 19 February 1907.

[29]Akiva Arie Weiss, *The Beginning of Tel Aviv* (Tel Aviv: Aynot, 1957), p. 66 [in Hebrew]; CZA, L2/71, Warburg to Ruppin, 6 November 1908; *Jerusalem*, vol. 1, Jaffa 1913, pp. 20–27 [in Hebrew]; A. Gabstein, 'Urban Building in Palestine', *Eretz Israel*, 7, 8 (1918), pp. 45–46 [in Russian]. Zaidman and Kark originally noticed the role of the German sociologist Franz Oppenheimer in introducing the term 'garden city' into the Zionist context: Miki Zaidman and Ruth Kark, 'The Genesis of Tel Aviv', *Zemanim*, 106 (2009), pp. 8–21 (p. 8) [in Hebrew].

[30]Willy Lange, *Land und Gartensiedlungen* (Leipzig: J. Weber, 1910), pp. 189–192.

than collected from rainwater and the streets were to be illuminated.[31] In addition to these general principles, there were also characteristics stemming from Zionist objectives, such as Hebrew schools and other important institutions providing public services, designated to be built at the heart of the suburb as a way of signifying their importance.[32] It should be noted that in Hampstead as well as in Howard's model, public buildings were located in the centre too, although usually as a matter of convenience.

It was likewise decided that the proposed inhabitants of the suburb should all be Jewish and it was forbidden to sell or rent houses to non-Jews, Muslims or Christians. The aspirations of the inhabitants to develop a society that reflected their Zionist-nationalist values determined this approach,[33] which somewhat corresponds to the ethno-racial segregation that was prevalent in other contemporary global colonial contexts. Indeed, the question of whether the Zionist movement constituted a colonial endeavour raises growing debate in Israeli academic circles.[34] Regarding the geographical location of the suburb vis-à-vis the old town of Jaffa, they decided to build it some distance away as an autonomous Jewish unit, stressing the development of community and national public life.[35] Though the then Jewish minority promoted a settlement policy, its garden city rhetoric and the accompanied modernist quality were distinct from the case of Dakar. One of the prominent differences was that the new suburb facing Jaffa was not identified with the official Ottoman regime and did not fit into its spatio-political agenda. In many respects, the new suburb even undermined the official regime in terms of independent and comprehensive administration and ideology – a regime that made difficulties for it from the start.

The garden suburb of Tel Aviv – initially called 'Ahuzat Bayit' (house or housing estate) – was built in accordance with the garden city programme. In March 1908, the Ahuzat Bayit Association purchased an area of 85,000 square metre about two kilometres distant from the centre of Jaffa in a north-easterly direction. It was sandy and almost useless for agriculture. The contract gave the Association an option to

[31]Alter Droyanov, *Tel Aviv Book* (Tel Aviv: The Book Committee, together with the Municipality, 1936), p. 75 [in Hebrew]; Weiss, *The Beginning*, p. 85; CZA, L2/578; Tel Aviv Municipality Archives (hereafter TAMA), the protocol file of the Ahuzat Bayit Association, 3 June 1907.

[32]TAMA, articles in *Tel Aviv*, vol. 2 (unpublished, in Hebrew); Smilanski, *With My Fellow*, p. 525; Weiss, *The Beginning*, pp. 115–123; *HaOlam*, 31 August 1909, p. 15 [in Hebrew]; Yossef Eliyahu Sheloush, *The Story of My Life* (Tel Aviv: published by the author, 1931), p. 139 (in Hebrew, reprinted by Bavel Publishers, Tel Aviv 2005).

[33]TAMA, the Prospectus of the Ahuzat Bayit Association, 31 July 1906; TAMA, the protocol of the meeting of 3 June 1907 [both in Hebrew]; Weiss, *The Beginning*, p. 85; Yechiel Tshlenow, *Five Years of Our Work in Eretz Israel (Palestine)* (Moscow: n.p., 1913), p. 55 [in Russian]; CZA, L18/105/4.

[34]See, for instance, the recent series of conferences, journal articles and books under the aegis of the Van Leer Institute in Jerusalem.

[35]CZA, L51/52, Ahuzat Bayit Association to the Jewish National Fund, 19 February 1907; TAMA, Ahuzat Bayit protocols file, protocol from the 3 June 1907; Droyanov, *Tel Aviv*, p. 84.

purchase additional areas bordering this stretch, thus securing the possibility of expanding the suburb in the future. Difficulties raised both by Arabs from Jaffa and Turkish authorities resulted in a year's delay, and the official purchase did not take place until March 1909.[36] The construction thus began in 1909, and by the end of that year some 50 homes had been built to house 500 people. In the middle of the suburb stood the Herzliya Gymnasium (the Hebrew High School), which remained for many years the centre of all Zionist-nationalist life in Tel Aviv. Yet in the final plan, there was a reduction in the extensive areas ideally allocated for parks and public gardens, since ignoring the financial constraints might have endangered the whole project (Figs. 9.5 and 9.6).[37] These physical features were opposed to the situation in colonial Dakar, where 'garden city' agenda fostered a generous allocation of land on the Plateau for a promenade, parks and gardens. This was enabled due to the colonial regime's monopoly over land ownership. By introducing

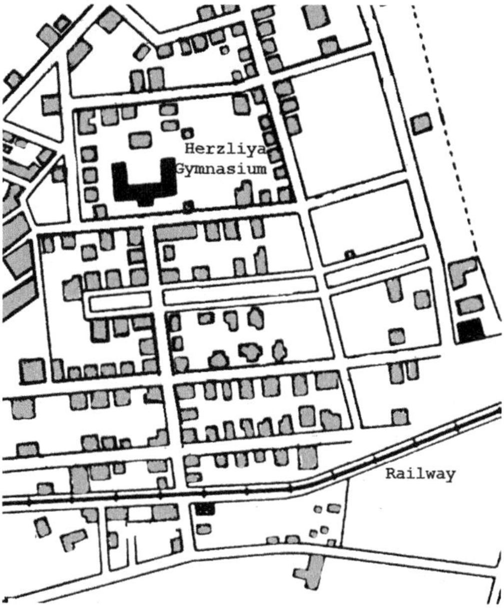

Fig. 9.5 Ahuzat Bayit's plan including the extent of the built space, 1918 (Authors' drawing after an image published by its establishing families)

[36]Droyanov, *Tel Aviv,* pp. 85, 92, 94; Smilanski, *With My Fellow*, p. 485; *HaOlam*, 4 May 1909, p. 15; Curt Nawratzki, *Die Juedische Kolonisation Palästinas* (Munich: Ernst Reinhardt Verlag, 1914), pp. 380–381; Sheloush, *The Story*, p. 131; TAMA, file of Ahuzat Bayit Association, minutes of meetings, 9–20 October 1907, 16 January 1908 to 9 November 1908, 9 January and 4 March 1909; CZA, L51/52, Letter from Levontin to Ruppin, 24 May 1909.

[37]TAMA, Unit 1, file 40, Letter from Treidel to Weiss, 8 January 1909; and Letter from Treidel to Weiss, 9 January 1909; TAMA, file of Ahuzat Bayit Association, minutes of meetings, general meeting, 17 and 20 February 1909; CZA, L2/633, Letter from Ruppin to the Keren Kayemet, 15 February 1909; *HaOlam*, 4 May 1909, p. 15.

Fig. 9.6 Panoramic westwards view of Ahuzat Bayit in the 1920s (Postcard in authors' collection)

new land laws and the nationalisation of land, all the obstacles that were originated in relevant Lébou customs, and Islamic and pre-Islamic practices, were removed.[38]

Nonetheless in Ahuzat Bayit, the inclusion of parks and public gardens within the urban sphere and the allocation of only 30 % of the building lot to the house itself represented a fresh approach to green spaces. It constituted the last word in urban planning, so as the borrowing of very term 'garden city.' This inspired even the Ottoman regime, which for centuries had been more interested in felling trees for various uses than in planting them. The latter embarked on various planting projects around the country in both urban and rural areas. One of the highlights in this Ottoman endeavour to improve the urban landscape in Palestine during the First World War, however, was the planting of 'Jerusalem Boulevards' in Jaffa in 1915. Oriented towards contemporary Tel Aviv (originally named Djamal Pascha Boulevard and later King George Boulevard), this street was 30 m wide and 800 m long. It was lined with Washingtonia trees that were tended by pupils of the Mikve-Yisrael agricultural school (Fig. 9.7).[39]

During its first years, however, Tel Aviv was a true garden suburb, with the town of Jaffa serving as its central city. A little later, however, Tel Aviv began to transform

[38] Alain Dubresson, *L'Espace Dakar-Rufisque en devenir* (Paris: ORSTOM, 1979), pp. 105–106; Assane Seck, *Dakar: Métropole ouest africaine* (Dakar: IFAN, 1970), pp. 122–131.

[39] Nili Liphschitz and Gideon Biger, *Green Dress for a Country: Afforestation in Eretz Israel, The First Hundred Years 1850–1950* (Jerusalem: KKL and Ariel, 2004), pp. 42, 47–49. See also Elli Schiller, *Jaffa and its Sites* (Jerusalem: Ariel, 1981), vol. 15 [in Hebrew].

Fig. 9.7 'Jerusalem Boulevards' in Jaffa today (Photo by Liora Bigon)

itself spontaneously into what would become a city. The large land reserves around the suburb, as well as the number of houses and their inhabitants, grew from year to year. By 1914, Tel Aviv covered 570,000 sq m, as opposed to 85,500 sq m in 1909, and the number of houses totalled 204 as opposed to 50 houses in 1909, with some 2000 residents. Tel Aviv began to take on various functions and commerce flourished there. Even prior to the First World War, there was a general feeling that the suburb was turning into a town.[40] Indeed, in 1921, Tel Aviv was officially proclaimed a local council and in 1934, a city, the first Jewish city in Palestine.

The success of the first Zionist-Jewish garden suburb in the proximity of a well-established old city was recognised even in 1910, and stimulated the establishment of similar garden suburbs in the Jaffa-Tel Aviv area and near other, older towns with large Jewish populations. Thus, public organisations and private initiatives were using the 'garden city' terminology undertaken with the aim of creating garden suburbs near Haifa on Mount Carmel, and in Tiberias, Safed, Gaza, Hebron, Ramla and Jerusalem. In the Arab sector, during the first quarter of the twentieth century, several elite neighbourhoods were established outside of the walls of the Old City of Jerusalem, though nothing of the kind existed in Jaffa. A few of them

[40]TAMA, Ahuzat Bayit Protocols File, protocols of meetings from the following days: 15 September 1909, 17 November 1909, 25 May 1910, 8 August 1910, 12 June 1912, 4 December 1913, 25 March 1913, 14 October 1913, 24 April 1914; *HaPoel Hatsair*, 1 November 1912, p. 21; *HaPoel Hatsair*, 26 December 1912, pp. 11–12; *HaZman*, 17 January 1914, p. 4; *HaOlam*, 5 February 1914; *HaHerut*, 27 May 1914, p. 1; *HaHerut*, 25 June 1914, p. 2 [all in Hebrew].

were sometimes called 'garden suburbs' and constituted an expression of modernist spirit amongst the Arab traditional elite – a spirit that was salient, inter alia, in the architecture of other Middle Eastern cities following the growing exposure to western influences from the turn of the century. But while the general appearance of these upper-class quarters was rather green, a direct or indirect reference to Howard's notions is ambiguous and left still untreated in the literature so far.

9.4 Conclusion

This chapter analyses 'garden city' rhetoric and practices in French Dakar and Tel Aviv (Ahuzat Bayit) in the late Ottoman period. In this, its contribution is twofold: first, it expands beyond the more common geography of the 'traditional' garden city historiography. This (anglophone) historiography has been preoccupied with developments in Britain itself, Europe and the global North. In our examination of the two case studies from the global South, which experienced colonialism, including an examination of a rich selection of primary and secondary sources from Senegal (officially francophone) and Israel (Hebrew sources referring to a distinctive urban phenomenon within the Jewish sector) – is in many respects enlightening. The second contribution of this chapter is conceptual: studies on the garden city treated the first developments in Britain as the exclusively legitimate source, while other modes of garden city application in other places in the West, and especially beyond the West, were treated as a distortion that did not always merit full understanding and recognition as equally legitimate. By bringing the two case studies from the background to the foreground of academic attention, this chapter aims at blurring the prevalent hierarchy in urban planning scholarship in which the West is conceived as the ultimate centre and the 'rest' as its ultimate 'margins.' In other words, it is argued that varieties and nuances of site-related garden city developments in the local and regional level should be conceived as an integral part of this globally disseminated phenomenon. This is without a value orientation of a global heart versus a global periphery. The term 'garden city' is therefore a generic term by definition, not specific.

As to French Dakar, the garden city rhetoric served the colonial administration in creating an attractive, prestigious and exclusive environment to house its employees, as well as an apparatus to attract them to serve in the (sub-Saharan) colonies. The Plateau exemplified a hierarchic space which responded to the rank and status within white society – a metropolitan element of social paternalism that accelerated and further crystallised in West Africa. The reference of the colonial administration to the term *cité-jardin* also contributed to residential segregation on a racial basis between the expatriate and the African communities, even if it was always informal in French sub-Saharan Africa. The modern, well-equipped, *pittoresque*, white residential areas within Dakar's federal ceremonious space only celebrated the colonial power. While the Dakarois were perhaps expected to imitate these residences, they never experienced similar planning developments as the 'dual cities' policy never provided for such planning amenities. Yet on the theoretical level, it might be illuminating to regard the concept of *cité-jardin* as it was implemented in French West Africa not just as a capricious and anti-social experimental realisation of indirect British planning ideas on colonial terrain. We shall rather consider the inherent flexibility of this notion (in Dakar, Tel Aviv and many other sites) as another important development in the history and historiography of the garden city.

As to Tel Aviv (Ahuzat Bayit), the process of the diffusion of garden city rhetoric was essentially transnational, with 'flows' arriving not directly from Britain to British Mandate Palestine, but much before that time, during the Ottoman regime through Zionist activists mainly in Germany. This rhetoric and its implementation in the form of garden suburbs embodied the Zionist model of urban settlement in Palestine, by serving the cultural and national aspirations of the Jewish sector. The result was, amongst other features, residential segregation of the two nationalities living in this territory, Jews and Arabs. Tel Aviv was the first garden suburb to be built in Palestine. Although it was planned as a suburb, it quickly developed into a budding city despite its proximity to Jaffa. The reasons for this were its unique location, and that the inhabitants of the new suburbs planned around Tel Aviv insisted that it should be a Jewish city, separate from neighbouring Jaffa. Tel Aviv, it seems, is the only garden suburb built under the influence of Howard's ideas ever to become an independent city. From the very beginning, the incorporation of garden city terminology into the urban development of Zionist settlement in Palestine was not merely morphological or aimed at the amelioration of living conditions, but an ideological matter. Garden city concepts offered more than an alternative to the industrial cities of Europe, the like of which did not as yet exist in Palestine (nor in colonial Senegal) and by thus provoked no reaction. The use of these concepts provided relief from unsuitable living conditions in the old towns after centuries of haphazard planning by the Ottoman regime, which has been considered chaotic.[41]

[41]For a different view, however, on the state of urban planning under the (late) Ottoman regime in Palestine, see Salim Tamari, 'Confessionalism and Public Space in Ottoman and Colonial Jerusalem', in Diane E. Davis and Nora Libertun de Duran (eds), *Cities and Sovereignty: Identity Politics in Urban Spaces* (Bloomington: Indiana University Press, 2011), pp. 59–82.

Chapter 10
Letters, Words, Worlds: the Naming of Soweto

Gordon H. Pirie

Abstract This chapter deals with the naming of South Africa's best-known black urban township, Soweto, and its subdivisions. At one level the presentation seeks only to detail the findings of an exercise in local toponymy. First, attention is given to the proposals put forward for labelling the southwestern townships, to the consultations and to the selection procedures followed. In the opening section it is the naming process which is spotlighted. The subsequent section switches the focus to the distinctive meanings and interpretations embedded in certain of the names submitted for consideration. This second section highlights the connotations which the southwestern townships acquired. Throughout the study, information is drawn from official archival records.

At a more general level of toponymical research a second objective of this chapter is to report some methodological lessons to be drawn from the case study. The archival material consulted reveals a rare glimpse of obstacles, agents, choices and procedures involved on an occasion when the naming of a place was a conscious and deliberative event. The example of Soweto provides a particularly sharp illustration of the proposition that naming is not only the outcome of private and impenetrable contemplation. In varying degrees the naming of places is a social activity. This is so either by virtue of it involving joint decision-making and/or in respect of it occurring within a given social milieu in which there are formal or informal conventions of name selection, assignment and adoption.

Keywords Soweto • Black urban townships • Toponymic procedures • Place-naming processes • South Africa

> Local names [...] are never mere arbitrary sounds devoid of meaning. They may always be regarded as records of the past inviting and rewarding a careful historical interpretation.[1]

---

[1]Isaac Taylor, *Words and Places: Or, Etymological Illustrations of History, Ethnology and Geography* (London: Macmillan, 1864), p. 19.

G.H. Pirie (✉)
African Centre for Cities, University of Cape Town, Cape Town, South Africa
e-mail: gordon.pirie@uct.ac.za

L. Bigon (eds.), *Place Names in Africa*, DOI 10.1007/978-3-319-32485-2_10

This chapter deals with the naming of South Africa's best-known black urban township, Soweto, and its subdivisions.[2] At one level the presentation seeks only to detail the findings of an exercise in local toponymy. First, attention is given to the proposals put forward for labelling the southwestern townships, to the consultations and to the selection procedures followed. In the opening section it is the naming process which is spotlighted. The subsequent section switches the focus to the distinctive meanings and interpretations embedded in certain of the names submitted for consideration. This second section highlights the connotations which the southwestern townships acquired. Throughout the study, information is drawn from official archival records.[3]

At a more general level of toponymical research a second objective of this chapter is to report some methodological lessons to be drawn from the case study. The archival material consulted reveals a rare glimpse of obstacles, agents, choices and procedures involved on an occasion when the naming of a place was a conscious and deliberative event. The example of Soweto provides a particularly sharp illustration of the proposition that naming is not only the outcome of private and impenetrable contemplation. In varying degrees the naming of places is a social activity. This is so either by virtue of it involving joint decision-making and/or in respect of it occurring within a given social milieu in which there are formal or informal conventions of name selection, assignment and adoption. Whereas the elucidation of these matters in the particular case of Soweto is the burden of the discussion which follows, the more general implications for the tradition of individualist and inferential toponymical studies are treated briefly.

10.1 The Naming Process

Concern over the naming of an urban Black settlement area to the southwest of Johannesburg first surfaced in the 1930s. At that time a start was being made on clearing inner-city slum-yards in terms of the 1923 Natives (Urban Areas) Act.[4] In the case of Johannesburg, slum-yard residents were removed to the farm Klipspruit

[2]This contribution was originally published under the same title in *African Studies*, 43, 1 (1984), pp. 43–51. Reprinted by permission on behalf of The Taylor & Francis Group Limited on behalf of the University of Witwatersrand. Minor stylistic changes and footnote clarifications were made by the Editor of this volume.

[3]Files 71/4/1 and A69/1 (2 vols each; 'Naming of Townships'), Archives of the Johannesburg City Council Non-European Affairs Department, West Rand Administration Board, Johannesburg.

[4]The use of the capital 'B' in 'black' throughout this chapter refers to the context of apartheid South Africa when the collective term (updating the more offensive nomenclature 'non-European') was used to designate everybody who was not classed as 'White.' The Population Registration Act of 1950 classified all the country's residents into Bantu (African), Coloured (mixed race), White and later Asians racial groupings. This Act had, however, been used and endorsed by earlier legislation such as the Land/Natives Urban Areas Acts passed between 1913 and the 1930s [The Editor].

Number 8 which had been purchased by the City Council in 1930.[5] At a meeting of the City Council's newly constituted Native Affairs Committee in September 1931, the suggestion was made that the settlement there be called 'Clarendon.' A counter-proposal from Councillor Edwin Orlando Leake advocated that the 'township' should bear "a native name."[6] There is no documentation of discussion on this matter. All that survives is the record of a motion adopted in February of the following year when it was agreed that the township be called 'Orlando.'[7] In a rather perverse decision the Council honoured the City's former Mayor and first Chairman of the Native Affairs Committee by using his name rather than his advice.

Adoption of the name 'Orlando' was not achieved without occasioning some difficulty. For instance, an objection was raised by the Postmaster-General to the use of 'Orlando' as a postal designation for the township. As he noted, there were several places in the world which bore the same name, such as settlements in Oklahoma, Kentucky and Virginia.[8] Responding, Mr Graham Ballenden, Manager of the Native Affairs Department, protested that missorting and confusion would be minimal as there were no other places of the same name in South Africa and because "at least ninety-nine percent of the correspondence addressed to the township will originate in South Africa." In a vague reference to the slim likelihood of locally addressed 'Orlando' mail going overseas, Ballenden added that "the possibility of confusion is extremely remote, particularly when the class of person is borne in mind."[9] One other difficulty concerned the railway authority which took the Postmaster-General's objection to raise uncertainty about what to call the township's railway siding. In time though it became clear that the name 'Orlando' would apply in some manner: "there is little prospect of the City Council changing its decision [...] whatever attitude the Postmaster-General takes up. No doubt, for postal purposes, some way will be found out of the difficulty."[10] In practice this meant using the name 'Orlando Native Township' as a postal address.[11]

[5]Pauline R. Morris and S. P. Silliers, *Soweto: A Review of Existing Conditions and Some Guidelines for Change* (Johannesburg: The Urban Foundation, 1980).

[6]Minutes, Native Affairs Committee, 2.9.31. In South Africa the term 'township' is a legal and physical construct. Outside of cadastral use the term is most commonly used in a racialised way to refer to suburban areas set aside exclusively for Black people. This usage differs from that in other colonial anglophone territories in sub-Saharan Africa, as established by Lord Lugard's Township Ordinance of 1917, first implemented in Northern and Southern Nigeria. According to this ordinance, the legal construct of a 'township' meant an enclave, or an area outside native administration and jurisdiction, that was separately governed and reserved for the residence of Europeans (and 'progressive' non-Europeans). Frederick D. Lugard, *The Dual Mandate in British Tropical Africa* (London: William Blackwood, 1922) [The Editor].

[7]Minutes, Johannesburg City Council, 23.2.32.

[8]Postmaster-General to Ballenden, 19.4.32.

[9]Ballenden to Postmaster-General, 22.6.32.

[10]Ballenden to Acting System Manager, South African Railways, 11.5.32.

[11]Acting Postmaster-General to Ballenden, 4.7.32.

In the early 1940s after the name 'Orlando' had been in use for some 13 years, there were plans made to extend Orlando township so as to accommodate the burgeoning Black population. It was proposed that the several extensions planned be named individually so as to "obviate the difficulty experienced in Orlando of identifying the locality where people live, both for postal and administrative purposes."[12] Commensurate with this idea, Mr L. I. Venables, then Manager of the Non-European Affairs Department (hereafter NEAD) approached a local European scholar for advice regarding the naming of extensions. Late in 1944 Professor Clement Martyn Doke of the Department of Bantu Studies in the University of the Witwatersrand, Johannesburg, submitted a list of several names, judging it "a very wise decision to use Bantu names in place of European."[13] Doke reviewed briefly the possibility of using names of well-known localities in 'Native Areas' (e.g., *Serowe* and *Dukuza)* and of employing great names from 'Native history' (e.g., *Sandile, Khama, Moshesh).* Respectively, he indicated that these two alternative conventions might create postal confusion and incite historical rivalries. A third option involved building compound names commencing in, for instance *Umzi* or *Motse*, followed by an element descriptive of the proposed locality (e.g., *Umzimithi*, "village of trees"). Against adopting this practice, Doke reasoned that it would produce long and similar words, and that it would present the difficulty of describing the townships before their development. The last hurdle was significant. Although not especially appealing, the featureless settings of extensions to Orlando township were not particularly distinctive in themselves. On the other hand, as emerged later, there were many names which could be used to anticipate a place of mass Black settlement.

Rejecting the three possibilities mentioned, Doke suggested that "the City Council would make a very fine gesture to the Non-European Section if they decided to honour some of the outstanding Bantu leaders of today who have done much for their people, are held in high esteem among them, and have done their best to co-operate with the Europeans in this country. Names could be chosen to represent the major tribes in the Union."[14] The names which Doke put forward for consideration were: *Dube* (Zulu; after Dr John Dube), *Jabavu* (Xhosa; after Professor Davidson Don Tengo Jabavu of Fort Hare College), *Mofolo* (Southern Sotho; after Thomas Mofolo, author of *Chaka), Moroka* (Tswana; after Dr James S. Moroka, prominent Free State medical practitioner and a future president of the

[12]Minutes, Non-European Affairs Committee (hereafter NEAC), 30.11.44.

[13]Doke to Venables, 28.11.44.

[14]Doke to Venables, 28.11.44. The colonialist derogatory term 'Bantu' (i.e., all 'Black' Africans) is used today only in linguistics, as an umbrella term referring to about 500 African languages. The word 'tribes' was used instead of 'ethnic groups.'

'Union' refers to the Union of South Africa (31 May 1910 to 31 May 1961) that was founded as a dominion of the British empire by unifying the four separated colonies of the Cape, Natal, Transvaal and Orange River. It came to an end when the country became a republic, under the present-day name of the 'Republic of South Africa' [The Editor].

African National Congress), *Makgatho* (Northern Sotho, a founder of the African National Congress), *Mqhayi* (Xhosa poet) and *Bokwe* (Xhosa minister of religion).[15]

After having been approved by the Non-European Affairs Committee (hereafter NEAC), the seven names on Doke's list were sent to the Postmaster-General for his comment. The suggestions were not entirely acceptable and the Postmaster-General asked that they be accompanied either by a prefix or suffix designative of place.[16] The matter was finally settled with agreement that addressing would involve placing the name 'Orlando' behind extension names, or using the convention 'P.O. Orlando.'[17]

Announcement of the names proposed for extensions to Orlando township drew some angry words from one of the Native Advisory Boards which was approached. Discussion of objections raised by members of the Western Native Township Advisory Board revealed bitterness that the University had been consulted rather than the Boards. One member indicated that "the names were not as genuine as the people would like them to be" and that "if the Advisory Board members had been requested to draw up a list of names they would probably have compiled it differently." Venturing beyond being affronted, another argued that "the names proposed were not worthy of the credit being paid to them and [...] that the naming of the townships should be left entirely to the Advisory Boards who would recommend names confined to Board members." A third member, Mr James S. Mpanza, moved that the proposed names be boycotted, and that the Boards deal directly with the Government on the naming issue. It was also suggested that the City Council be asked to withdraw their recommendations to the Postmaster-General. The compromise which was reached saw the naming matter being referred to the Advisory Boards. Their recommendations as to suitable names were to be returned in time for the next meeting of the Joint Advisory Boards.[18]

Notwithstanding agreement on the urgency of finding acceptable names, no alternatives were lodged by the time stipulated. A stalemate ensued, with one Advisory Board member persisting in his view that Boards "were entitled to the naming of the townships and would deal with the matter as they considered fit."[19] In the absence of documentation concerning the City Council's response to the deadlock, it may be inferred that the Council opted unilaterally for the convenient interim solution of naming extensions according to compass bearing and sequence (e.g., Orlando West, Orlando East, Mofolo North, Mofolo South, Moroka North, Central West Jabavu and Orlando West No. 2). At least this is the impression given by a notable incident toward the close of 1946 when the Orlando Native Advisory

[15]Doke to Venables, 28.11.44.

[16]Venables to Postmaster-General, 6.2.45.

[17]Postmaster-General to Venables, 28.2.45.

[18]Minutes, Joint Advisory Boards, with Manager, Non-European Affairs Department (hereafter NEAD), JAB/NEAD, 11.4.45.

[19]Minutes, JAB/NEAD, 16.5.45.

Board recommended that Orlando West No. 2 be named 'Mpanzaville.'[20] According to Secretary W. S. Pela, the name of James Sofasonke Mpanza (a Board member) was brought forward "not because of any outstanding work which the latter had done for the residents of Orlando, but purely and simply out of spite and to cause embarrassment to the Council"[21]; it was Mpanza who, in the face of chronic under-provision of Council housing, had organised a squatter movement on a municipally owned site adjoining Orlando East. This massive squatting movement galvanised the Council into erecting temporary 'breezeblock' shelters in an area which came to be called 'Shantytown.' By early 1947 Mpanza effectively led a movement of some 60,000 squatters. In exchange for providing residents with hessian and guards, and for selling trading rights, Mpanza collected fees and levies.[22]

At least in part, Pela's objection to the name 'Mpanzaville' stemmed from his suspicion about Mpanza's motives, from his view that Mpanza exploited local residents, and from his dislike of the shrewd tactics Mpanza used to gather support for the use of his name. Prior to the public meeting which the Orlando Advisory Board called in February to consider the new name, Mpanza organised meetings at which he reportedly:

> told the people of the Shelters that he was going to give them those white [breeze-block] houses and that the name for that township was now being sought and that if they did not attend the meeting [. . .] they would not get those [breeze-block] houses and that they would forever remain in those unhygienic houses [the shanties]. The result was that the whole of Shantytown flocked to the meeting to show their solidarity and confidence in their leader.[23]

An estimated 1200 residents attended the meeting. Despite being told that "the new place at Orlando West No. 2 was going to house other people from other places and not known by Mr Mpanza nor knowing him", the name 'Mpanzaville' was unanimously endorsed.[24] Fearing that implementation of the resolution would provoke "resentment and bitterness among the responsible residents of Orlando", Pela, regarded by one official as "one of the most prominent and respectable Native tenants of Orlando",[25] proposed that the name of the new extension honour one of four other prominent people. The four, all of whom had been or were still associated with the African National Congress, were Mr Sefako Makgatho (past President-General), Mr Richard Victor Selope Thema (Co-founder, Member of the Natives Representative Council and Editor of *Bantu World)*, Mr D. S. Letanka (past Vice-President of the Transvaal Congress), and Mr S. P. Matseke (past President of the

[20]W. S. Pela, Secretary, Orlando Native Advisory Board, to Venables, 5.12.46.

[21]Pela to Venables, 21.3.47

[22]For a detailed treatment of Mpanza's movement see Alfred William Stadler, 'Birds in the Cornfield: Squatter Movements in Johannesburg, 1944–1947', *Journal of Southern African Studies* 6, 1 (1979), pp. 93–123.

[23]Pela to Venables, 21.3.47.

[24]Reports of public meeting, Communal Hall, Orlando, 27.2.47; Secretary, Orlando Advisory Board and Senior Superintendent, Orlando Township.

[25]W. J. P. Carr, Deputy Manager NEAD, to NEAC, 3.7.47.

Transvaal Congress). Pela also mooted the idea that some of the Council's townships be named after leading African chiefs such as Sekhuhkuni.[26]

Whatever the role played by Mpanza in Orlando, the suitability of using his name in or as that of the new township did not commend itself to the controlling body in the City Council. His antics fell foul of the informal 'principle' enunciated by Doke, namely, that the Council should honour local leaders. Furthermore, the opinion had already been expressed that it was "more acceptable to name national figures rather than local figures."[27] Reports of Mpanza's "usual extravagantly wild manner [...] and shouting" at a later, disorderly public meeting in Orlando at which senior officials of the NEAD were present did not help his case.[28] Noting that "it is obvious that if Orlando West were officially named 'Mpanzaville' it would make the task of administering that section of the Location extremely difficult, if not impossible", Venables sought advice from Johannesburg's Chief Native Commissioner.[29] The affair ended abruptly with the latter's endorsement that the Council should not use the name 'Mpanzaville.' In a dismissive rider, the Chief Native Commissioner saw fit to add that "in any event the name 'Mpanzaville' does not appeal to me as being suitable for the naming of a native location."[30]

Slighted by the Council's refusal to accept what it regarded as the majority's preference for a new township name, the Orlando Advisory Board resolved to adjourn indefinitely. Undeterred by previous criticism that Black participation in naming was confined to selection from a pre-ordained list, the Council again approached Doke for advice. On this second occasion the Professor drew up a list comprising the names *Makgatho, Plaatje, Msane, Thema* and *Rubusana.* [31] The Council sent these names to the Orlando Advisory Board with Thema's name at the head and Makgatho's at the foot. Accompanying the list was a statement to the effect that the name 'Thema' would be selected by default if the Board refused to co-operate in the review process.[32] Objecting strongly to the names submitted, the Orlando Board sent a delegation of protest to the City Council. In response to a charge of undermining the wishes of a majority of Orlando residents, there came the retort that "the naming of European and Non-European townships was at the discretion of the Council" and that after all "a street in Orlando had been named after Mr Mpanza." The courtesy extended in that gesture was queried though by delegate O. S. Mooki who referred to "much discontent over the naming of those streets, as well-known people had lanes named after them and others not nearly so well-known were given the honour of large streets bearing their name."[33] The

[26]Pela to Venables, 21.3.47.

[27]Minutes, JAB/NEAD, 11.4.45.

[28]Carr's memorandum on a public meeting, 23.3.47.

[29]Venables to Chief Native Commissioner, 26.3.47.

[30]Chief Native Commissioner to Venables, 3.5.47.

[31]Doke to Venables, 3.6.47.

[32]Minutes, NEAC, 3.7.47.

[33]Minutes, NEAC, 12.11.47.

deputation from Orlando withdrew and notified the Council that it was appealing to the Union's Minister for Native Affairs for redress on the naming of Orlando West No. 2.[34]

After the acrimony and impasse of the late 1940s the naming of other southwestern townships of Johannesburg proceeded smoothly. In the early 1950s names were sought for many new townships. The exact process of consultation and adoption is not documented in the sources used here,[35] but it seems that Advisory Boards tendered the names *Moletsane* (Conqueror-chief of the Bataung in the 1830s), *Chiawelo* (Venda; "place of rest") and *Phiri* (Sotho; "hyena", and Pedi family name). Mr W. J. P. Carr, by then Manager of the NEAD, was apparently responsible for suggesting the names *Molapo* (Sotho sub-group), *Dhlamini* (Nguni family name), *Senaoane* (City Council sports organiser), *Mapetla* (Sotho family name) and *Zondi* (Nguni family name).[36] The names of other townships, *Jabulani* (Nguni; "rejoice"), *Naledi* (Sotho; "star"), *Zola* (Nguni; "be calm") and *Tladi* (Sotho; "lightning") "were all suggested by the residents, most of whom were moved from squatter camps."[37]

In 1956 the announcement by the City Council of a massive £3 million housing initiative in southwestern districts (Sir Ernest Oppenheimer was a key figure in negotiating the mining house loan) launched a new phase of township-naming history. After some discussion it was decided that a collective name should be found for the array of individually named townships. The model was that of a city and its several suburbs. At first, suggestions were simply invited from staff in the NEAD and in the Housing Division of the Council. By early May 1957, eighteen employees had responded to the incentive of a prize by submitting thirty names for consideration.[38] Unattracted to any of the proposals, the search continued for an overall name for the dormitory suburbs which the Council had taken to labelling the 'South Western Native Areas.' Carr himself offered a prize of £10 for the suggestion of a suitable name. By August 1958, 114 entries had been received. None were judged appropriate. A wider imagination was sought and the 'competition' was opened to the public by way of Press announcements.[39] Letters flooded in from blacks and whites living both near and far afield. Although the word-lists were soon stretched by a further 425 names, in February 1959 Carr declared the results "disappointing in that no name has been suggested which is strikingly suitable."[40] "In the absence of anything better"[41] however, Carr shortlisted 11 names from those put forward. In the order of his preference these were: *Goldella Serena* "may

[34]O. S. D. Mooki, Secretary, Orlando Advisory Board, to Venables, 25.11.47.

[35]See endnote 2.

[36]Annotated letter, Carr to Johannesburg City Librarian, 15.2.61.

[37]Carr to Town Clerk, 25.11.60.

[38]Carr to Councillor H. Miller, 7.5.57.

[39]*Sunday Times*, 31 August 1958; *Weekblad*, 14 November 1958.

[40]Draft report to NEAC, 10.2.59.

[41]Draft report to NEAC, 10.2.59.

peace prevail"; *Sothuni* derived from 'Sotho' and 'Nguni'; *Dumuzweni* "famous the world over"; *Ulundi* "name of a mountain"; *Entabene* "on the mountain"; *Kwantu* "at the Bantu's place"; *Umuzi* "homestead"; *Ukwezi* "morning star"; and *Umuzi-Kazi* "a very big city."[42] Even after compilation of the short-list, names continued to be submitted; there were a further 137 by March 1959. Concerned about the preponderance of Nguni names on the short-list, Carr appointed a subcommittee of the Joint Advisory Board to consider the matter. In what must surely have been a political move, Mr Mpanza was appointed representative for Orlando on the six-man body.[43]

Considering their brief, the subcommittee's recommendation must have come as something of a surprise to Carr, who by this time had firmly established himself as the key official in the naming process. In a letter of October 1959 the naming subcommittee reported that the *only* suitable names for the South Western Native Townships were *Kwa Dulaza* "Dulaza's place", *Thari' Ntshu* "black baby's carrying-sling" or *Kwa Mpanza* "Mpanza's place."[44] There was no reaction to these ideas for 4 months. In the interim a reminder in the Press of the naming 'competition' brought a further rash of names.[45] In February 1960 Carr declared that two of the three names suggested by the subcommittee did not measure up to the naming conventions established by the Union Native Affairs Department[46] and that *Thari' Ntshu* was "also unsuitable as Europeans would find it difficult to pronounce."[47] In retrospect this appears a rather lame apology for Anglo-Saxon linguistic overlording. Again the decision was deferred. A fresh start was made, this time by asking individual Advisory Boards to submit three names for consideration. In the course of their deliberations further 'competition' entries deluged some 270 more words on to desks in the NEAD. The recommendations made by eight Boards (Dube and Chiawelo made no returns) contained not a single common name though the selections made repeated reference to the notion of a large settlement.[48]

What became of the Boards' suggestions is uncertain. More appeals for names were made through the Press during 1961. At the beginning of September more than 400 names made up the new word-lists. For the first time the word 'Soweto' had been suggested,[49] being a straightforward combination of the first two letters in the words South Western Townships. Hand-written remarks at the foot of the listing compiled in the NEAD in early September show a preference among two departmental officials for either the name 'Soweto' or 'Toweso', with a third official opting for 'Randala' *(rand* "to earn"; *lala* "to relax, sleep or rest"). The

[42]Draft report to NEAC, 10.2.59.

[43]Minutes, JAB/NEAD, 1.7.59.

[44]J. M. Mamakoko, Secretary, Naming Subcommittee, to Carr, 14.10.59.

[45]*Sunday Times*, 1 November 1959.

[46]Circular, 28.10.57.

[47]Minutes, special NEAC meeting, 4.2.60.

[48]Notes, NEAC meeting, 18.5.60.

[49]Mrs C. Stanissis of Pretoria to Carr, 8.6.61. Prize of R20 awarded to Stanissis 30.10.63.

names ‘Soweto’ and ‘Toweso’ were made items on an agenda for a Joint Advisory Board meeting, with Carr noting that “both names are ‘coined’ [. . .] short, easily pronounceable and do not favour one of the main language groups.”[50]

Among the names presented for consideration towards the end of 1960 was one which appeared in a reader’s letter to a newspaper editor[51] but which did not ever get on to the NEAD lists. The suggestion was ‘Magebula’ (Zulu; “one who cuts off a chunk”). The name was intended to honour Mr Mpanza who, as already discussed, had commandeered a piece of municipal land for his squatters’ camp. Responding to this suggestion, one Orlando resident warned that whereas “leading hundreds of poor, cold, hungry people of different races to the wilderness was indeed a courageous undertaking”, the name ‘Magebula’ would eulogise a recalcitrant and illegal act by someone who was accepted as leader by only a small minority in a small part of the township complex.[52] The possibility of somehow honouring the name of James Sofasonke Mpanza was not to surface again for a decade. Then, in 1970, Soweto’s Urban Bantu Council resolved that in future one of the extensions to the existing townships would be named ‘Mpanzaville.’[53] More recently, in August 1983, the Soweto Council unanimously turned down a motion to rename Soweto after Mpanza, but agreed to rename Orlando East ‘KwaMpanza’ (Mpanza’s place).[54]

No decision was reached at the Joint Advisory Board meeting at which the names *Soweto* and *Toweso* were studied. A second naming subcommittee was duly appointed to reconsider the entire issue. Mr Mpanza was again made a member. Reporting in November 1962, each member of the subcommittee put forward one name on a list “from which the Council can make its choice”: *Soweto, Sawesko, Swestown, Phaphama Villages* and *Partheid Townships* (Mpanza).[55]

During the following December and January, Carr made it plain that he was toying with the names ‘Soweto’ and ‘Sawesko’, preferring the former.[56] Acting for the Council, Carr had been given *carte blanche* by the naming subcommittee. The Press gave strong indications that, at last, matters were coming to a head and that, failing superior suggestions, the name chosen would indeed be ‘Soweto.’[57] As if recoiling at that contrivance, the public submitted some 300 new names within three weeks of the Press announcement. By this stage too, the favour shown ‘Soweto’ induced a spate of anagram building [e.g., *Sowetu, Sowantu, Sotewe, Sawata, Sewato, Jobatu, Josowe, Sorana* (South Rand Native Areas); *Sowerana, Rabanto* (Rand Bantu Townships); *Esoswato* (Sir Ernest Oppenheimer South Western Areas

[50]Carr to Administrative Assistant, NEAD, 1.5.62.

[51]*Star*, 3 November 1960.

[52]P. Mashego to Carr, 24.11.60.

[53]Secretary, Urban Bantu Council, to Carr, 5.11.70.

[54]*City Press*, 28 August 1983.

[55]F.S.M. Mncube, Secretary, Naming Subcommittee, to Carr, 5.11.70.

[56]Circular to NEAC, 11.12.62.

[57]*World*, 15 January 1963; *Star*, 18 January 1963.

Township); *Seoncia* (Sir Ernest Oppenheimer Native City in Africa); *Penema* (corruption of Oppenheimer); and even *Doityose Township* (Do it Yourself)!]. An initial attempt to resolve the naming issue was made at an NEAC meeting on 14 March 1963. By this time some had begun to doubt both the possibility and wisdom of finding a composite label. A vote was taken first on an amendment "that the South Western Bantu Townships retain their present name." The amendment was defeated and the main motion was carried: "that the South Western Bantu Townships be given a composite name." Voting was tied though on the formal motion "that the South Western Bantu Townships be named Soweto." Whereas his casting vote had broken an equality of votes on both the amendment and first motion, the Chairman declined to invoke the same privilege "a third time and have done with the seemingly irresolute nightmare there and then."[58] Instead, the naming-issue dragged on for yet another month before the name 'Soweto' was approved at the April NEAC meeting. The fact that the Advisory Boards and that the NEAC had both approved the name was taken as sufficient indication that the name should be adopted; never at any stage was anything like a referendum contemplated. The City Council acceded to the NEAC's April resolution and, as required, referred the name 'Soweto' to the Minister of Education, Arts and Science for his approval.[59] Without giving reasons the Minister withheld consent.[60] Untouched, the Chairman of the NEAC remarked that Ministerial disapproval "does not alter our decision. We submitted the name [...] as a courtesy"[61]; "there is no legal obligation to get the name approved and we will continue to use it."[62] Official dislike of 'Soweto' or no, it was reported that "thousands already like the name and use it, and it will be hard to make them stop."[63]

The decision to adopt an inoffensive but bland sequence of letters for a collective township name was not welcomed in all quarters. This was the more so as, for the duration of the public 'competition', the Press had highlighted the many euphonious, exuberant, romantic, mellifluous and apt suggestions.[64] Three objections made by members of the public were that the name was "similar to Russian Communist Soviet Republic", that it had "the flavour of a telegraphic address", and that "we want a Bantu name not a direction."[65] Even the innocuous and synthetic amalgam 'Soweto' did not sidestep some overtones which the City

[58]Minutes, NEAC, 14.3.63.

[59]Town Clerk to Carr, 3.5.63.

[60]Secretary, Department of Education, Arts and Science, to Town Clerk, 17.6.63.

[61]*Star*, 20 August 1963.

[62]*Rand Daily Mail*, 9 August 1963.

[63]*World*, 8 August 1963.

[64]For example, *Star*, 22 October 1958.

[65]The latter objection is particularly ironic in view of Nelson Mandela himself expressing dissatisfaction with how the name 'South Africa' referenced a direction more than a place. This is according to a personal communication with Arthur Goldreich, who was jailed along with Mandela and others in 1956 during the anti-apartheid struggle [The Editor, Goldreich interviewed in Herzliya, Israel, in February 2010].

Council must have regarded as unfortunate. "Pronounced by some Africans, the word sounds remarkably like 'so where to?' – a not inappropriate name for a locality whose inhabitants as 'temporary sojourners' have a distinctly uncertain future. With their close experience of 'endorsing out' [. . .] they may find Soweto appeals to their sense of irony in this land of 'forever hamba.'"[66]

The preceding discussion of the naming of Johannesburg's southwestern townships has stressed the mechanics and politics of arriving at names for individual townships and for the entire community. In the process several public bodies and many public servants were involved, including the Postmaster-General, the railway authority, a Cabinet Minister, the NEAD, its Committee and its Manager, township Advisory Boards and the umbrella Joint Advisory Board. In addition the Press and the public played a part in the proceedings. Solicitation procedures, and consultative roles varied and produced the inevitable bitterness and stalemates, following which arbitrary names were used to designate township extensions. Once new townships had been named individually, attention turned to devising a collective name. Over a period of 6 years more than 1000 alternatives were mooted. Their suitability was tested loosely against criteria of length, pronounceability, tribal favouritism and screened according to sonorous and visual appeal. There was apparently no consideration given to the way in which the name would be reworked to describe township residents (Sowetans) and township attributes and ways (Sowetan). Overall, township naming was clearly very different from what is generally assumed to be the gradual, almost unnoticed way in which place-names become accepted lore.

10.2 The Meaning of Names

The sheer bulk of the letters and words written in connection with township naming precludes an exhaustive recapitulation of the lists and the often entertaining explanations which accompanied the suggestions. One 'competitor' put forward the name *Duba Duba*, "which incidentally is my personal name and a very good one too"! Some letters were crudely patronising. Punctuating her letter with the remark that naming African townships should not be taken too seriously, one writer had it (curiously) that "*Buriton City* will do [. . .] it is an easy name for them to remember." Besides these interesting reflections on participants in the naming competition, the mass of correspondence containing proposals also affords some

[66]*Rand Daily Mail*, 21 August 1963. While in Indonesia the word 'hamba' means 'a slave', in the South African context its direct translation would be something like 'go away'. The phrase is poignant as Africans were always being told to go away, or being sent away. Originally a Zulu word, the tone of voice, facial expression and context establish its meaning: said with a smile it would mean 'go on, run along now!'; said angrily, it would mean 'fuck off, go hang.' The addition of one word (namely, 'kahle') changes hamba into 'go well', which is a common, warm and respectful farewell greeting. In our case 'forever hamba' is quite a dismissive remark as in 'get out of my sight forever'.

insight into the connotations which the southwestern townships sparked in South African minds. Places (perhaps unseen ones particularly) conjure up mental images; the circumstances of their settlement strike chords; words echo imagined worlds. In short, inspection of even just a sample of the many suggestions throws light on how South Africans regarded the growing nucleation of match-box houses and rudimentary infrastructure on the outskirts of the 'White' metropolis.

Apart from anagrams, adaptations of well-known words (e.g., *Bantuville, Egoligauteng, Proletaria,* and *Arbeidorp*) and allusions to prominent people (e.g., *Dingaansville, Verwoerdville, Oppenheimer, Bantu Eldorado,* and *Carrsrust*),the names put forward suggest a spectrum of conceptions of place which range from the concrete to the abstract. For some the townships acquired significance only as a material stock of buildings sheltering vast numbers of 'Black' people in a specific position relative to the 'White' city. Some of the names which reflect this conception are *Setletle* "colossal village", *Shimita Miti* "a monster that swallows villages", *Mzimkulu* "great place", *Emnyameni* "black spot", *Mzimnyama* "dark city", *Kwantu* "at the Bantu's place", *Kayazizwe* "home of the tribes", *Egolinyana* "little Johannesburg", *Boro-Phima-Kaya* "south western township", *Eshonalanga* "western", *Phala Borwa* "best of all in the south", and *Vergenoeg* "far enough." The last five names in this list are examples of the relatively few suggestions which were 'positional' and which were specific to particulars of the townships in question. Only one suggestion was ever made that the southwestern townships take the name originally used by Blacks to denote the reeded stretch of land between New Canada and Randfontein: *Motlhakeng* (Sotho) or *Mhlangeni* (Nguni). These names were used "all the years even before the locations were built on this area."[67]

A second major category of names submitted made reference to the southwestern townships as a destination, home, for a large number of people: *Valhalla* "journey's end", *Zamlandela* "everybody follows him", *Inqaba* "place of refuge", *Bosa Khuthleng* "everlasting home", *Kholumolumo* "swallowed up all people within", *Khayalam* "my home", and *Kwakiti* "our place." Numerous suggestions were put forward in a third category in which the townships were depicted as places in which there would be good fortune and in which hopes and ambitions would be realised. A small selection includes *Ebhongweni* "place of ambition", *Endwneni* "place of fame", *Tswelopele* "progress", *Etembeni* "where there is hope", *Thembalihle* "good hope", *Boiketlo* "relaxation", *Tokolohong* "freedom" and *Zweledinga* "land of need." The Hebrew word *Teshuba* "reorientation" was an unusual submission made by a Moroka resident. Not dissimilar in their abstract references and emotive expression were names alluding to rest, peace, happiness and goodwill: *Boitumelong* "where people are happy", *Phomolong* "place of rest", *Boipelo Township* "township of joy", *Kagisano* "place to live peacefully", *Jabula* "be happy", *Ezibeleni* "kindness in abundance", *Tsakani* "rejoice", *Khotso* "peace", *Kwa Nonleze* "place of mercy", *Pelankani* "finish strife."

---

[67]B.J. Mabuza to Carr, 31.1.63.

The grouping of names in the manner done here unveils interesting cognitive elements in the many meaningful proposals which were rejected. That the translations which accompanied the proposals were necessary at all speaks to the control which White English speakers had over the entire naming exercise for three decades. That the translations may fail to capture subtle meaning is suggested by the difficulty in interpreting the significance of names such as *Isibakabaka* "sky", *Lwandle* "sea", *Mochochonono* "comet", *Bua-Feta* "speak and pass" and *Gezubuso* "face wash." A notable omission from the names discussed is the large batch of English language words offered for consideration. By and large their referents are concrete and straightforward covering an astonishing range: *Afric-holme, Goodville, Bordertown, New Canaan, Necessity Kloof, Needles, Safric, Dreamsite, Craig-Royston, Copper Mine, Prince Charles, Coon's Kraal, Sputnik, Darkiesuburban, Blackpool, Segregaria, Black Birds Bunk*, and *Creamland.*

10.3 Conclusion

One century and a half after Taylor composed the lines cited at the head of this chapter, the naming of Soweto would seem to have given lie to his thesis concerning local place-names and concocted sounds. At first sight too, the simple etymological roots of the word 'Soweto' do not commend its analysis as an historical signpost. Both conclusions are false. As the first section of the preceding presentation sought to show, even contrived place-names may have a fascinating history and may themselves be pungent statements of socio-political conditions and practices of earlier times. Over the few decades during which the word 'Soweto' has been in use it has become known throughout the world to a degree which other place-names (e.g., *Dumuzweni* (Nguni) and *Tumamafatseng* (Sotho) "famous the world round") might never have made possible. That 'Soweto' has no nicknames or abbreviations testifies to its ready acceptance.[68] Despite its terseness and linguistic pallor it has acquired a measure of international infamy. Paradoxically, a word as empty of etymological meaning as 'Soweto' has a distinctiveness which has made it an encyclopaedia of meanings or associations. There being no risk of confusing the name with that of any other place, 'Soweto' is immortalised as the archetypal 'Black' urban dormitory in apartheid South Africa. Despite its fabricated blandness there is indeed a world in the letters of the word 'Soweto.'

[68]This observation is outstanding in the light of the fact that in urban Africa – similar to many places especially in the global South – (post-)colonial situations created a multi-layered toponymic reality (see Introduction). This is aside from the current South African urban context, where post-apartheid government's efforts to get rid of the apartheid legacy have resulted in some places going by more than one name, used at the same time. Serena Coetzee and Antony Cooper, 'The Value of Addresses to the Economy, Society and Governance: A South African Perspective', paper presented at the 45th Annual Conference of the Urban and Regional Information Systems Association (URISA), 20–23 August 2007, Washington, DC, USA [The Editor].

In amplifying the importance of studying the social context of naming places it may be said in conclusion that the naming of Soweto provides an excellent demonstration of the necessity of modifying the inferential and individualist paradigm of toponymical research. Typically, etymological inquiry is used to uncover patterns of naming and these regularities are treated as beacons for inferring landscape and settlement histories. Whereas this procedure does give attention to the linguistic roots of place-names it fails to consider their social origins and the social settings in which they become household words and acquire significance. Naming is not some mysterious auto-suggestive event, however, and this is so no matter how fierce the battle being commemorated, how inspiring the landscape, how charismatic the first settler family. Whether permissive or restrictive there are tacit or explicit rules in society about who has the authority to select and apply names, what names are appropriate and whether an assignment will be adopted. In the particular case of Soweto the tortuous institutionalised procedure of naming makes plain the importance of having toponymical study penetrate beyond individualism to concern itself also with the social relations of deference and defiance upon which naming is contingent. This perspective on naming makes problematic too the unanimity of name selection which seems encrusted in the singularity of a place-name. It is unfortunate that there will be only a few instances in which place-name alternatives will have been recorded as testimony to the social nature of the activity. As was indicated in the second part of this chapter, names which are suggested but which never come into use are themselves rich repositories of information about the perception of places.

Chapter 11
South African Identity as Reflected by its Toponymic Tapestry

Barbara Meiring

Abstract The toponymic landscape of South Africa, as colourful as its flag, a kaleidoscope of colour, reflects the diversity of South African citizens. This chapter aims to point out that the geographical names, in this case referring to urban entities, collectively also reflect the hearts and minds of those who gave the names, ultimately expressing what it means to be a South African in the face of shared historical and cultural experiences that motivated the bestowal of the name. The effect of the policy of transformation will be discussed focusing on how transformation differs from mere change and how these two concepts apply to the geographical names of South Africa and its peoples. Using the onomastic framework of the etymological and associative meanings of place names, it will be pointed out that being South African involves both shared and differentiating experiences leading to the conclusion that there is no definition for a South African identity and that a national identity is subject to a consciousness and tolerance of social diversity and need.

Keywords Toponymic landscape • South-African heritage • Political transformation • Multicultural society • Naming etymology • Linguistic variety • Ethnic variety

11.1 Introduction

The premise that a South African spirit and the descent, history, cultural and economical background of the country's peoples can be derived from geographical names rest on the onomastic principle that place names have meanings that reflect the motivational forces behind the bestowal of the names.[1] The standardisation

[1]Another version of this contribution was originally published under the same title in *Acta Academica*, 44, 3 (2012), pp. 24–51. This version was rewritten by permission on behalf of The University of the Free State, South Africa.

B. Meiring (✉)
South African Geographical Names Council, University of South Africa, P.O.Box 151, Boesmansriviermond 6190, Eastern Cape, South Africa
e-mail: meiriba@vodamail.co.za

L. Bigon (eds.), *Place Names in Africa*, DOI 10.1007/978-3-319-32485-2_11

and official acceptance of the name by an authoritative body comes later but its inherent meaning stays the same even if the motivation is not relevant any more, making the meaning opaque. Not all names are descriptive in which cases an associative, emotional or etymological meaning indicates what the people of the land encountered, what they value and how history influenced their lives.

Following Edward Said this study does not accept the politics of identity whereby sanctioned narratives define who you are, who your heroes are, what you should value, etc. by internalised norms of self-identity.[2] Nor is this point of view a disregard for the cultural heritage of the various cultural groups in South Africa. In her book *A Change of Tongue*, the South African author Antjie Krog touches on the psyche of being a South African.[3] There is no question that her portrayal of various South Africans in clear language cut right to the core of motivations and intentions, exposing the essence of historical and social incidents. The moulding of a South African nation in the aftermath of the liberation struggle from 1994 onwards is a dynamic process involving change for everyone to attain an outcome of transformation. Krog's quotation of Noam Chomsky is relevant not only to her interpretation of transformation, but also to this chapter on toponyms as a reflection of being South African in the process of transformation:

> Some rules are transformational: that is, they change one structure into another according to such prescribed conventions as moving, inserting, deleting, and replacing items. Transformational Grammar has stipulated two levels of syntactic structure: deep structure (an abstract underlying structure that incorporates all the syntactic information required for the interpretation of a given sentence) and surface structure (a structure that incorporates all the syntactic features of a sentence required to convert the sentence into a spoken or written version). Transformation links deep structure with surface structure.[4]

Although Chomsky was talking about Transformational Grammar, it is also applicable to other fields of study as the basic assumption stays the same, i.e., some changes are on the surface while the deeper changes imply features that are linked to the visible surface changes by a process called transformation. Some things can therefore change on the surface without being transformation, if there is no deep internal change. Applying this to the South African toponymic scene we find that the change of dominant political forces led to inversions of the toponymical landscape, giving a diachronic picture of the various language and cultural groups who visited a particular place. The question is whether these were surface changes or whether there was also a change in the internal composition and activities of the relevant communities after the name change. In many cases where geographical names are proposed to be changed motivated by negative connotations experienced by one sector of a community, another group has more positive feelings regarding the name. Although this often points to the associative meaning of names, the

[2]Edward Said, *Culture and Imperialism* (London: Vintage Books, 1994), p. 380.

[3]Antjie Krog, A *Change of Tongue* (Cape Town: Struik Publishers, 2003).

[4]Krog, A *Change of Tongue*, p. 5.

etymological and lexical meaning of some names also point to human involvement that could lead to either positive or negative reactions.

Explaining how the meanings of different geographical names reflect the experiences and feelings of man, the name giver implies a search for the conditions as the motivating force under which the names were given. Common traits of being South African are illustrated by approaching naming as a linguistic act, a product of the human mind. An assessment of the extent to which the changing of names reflects true transformation of society and the demand for social cohesion in the light of political and social changes, urban renewal and renaming will be discussed in the light of the linguistic and cultural variety of South Africans enhancing the toponymic landscape, giving it depth and immeasurable historical value.

11.2 Methodology and Data

Focusing on urban toponyms most of the data used for this chapter is a selection of names of cities, towns, townships and settlements as well as the new names of Local and District Municipalities and Trans-Frontier Parks given after 1996 – from Peter Raper's *New Dictionary,*[5] data lists from the governmental Department of Arts and Culture and various brochures in the tourism industry. The presentation and processing of the data is based on an interpretative framework with an explanatory rather than a descriptive approach.[6] These names include most of the languages of the country with the African languages collectively being the majority. All aspects of meaning, be it etymological, associative, referential, descriptive or lexical, form the descriptive backing of a name and are therefore the basis for further interpretation with regard to the motives of the people involved in the name giving process. The descriptive backing of names is often hidden and only through research do the associative meanings of a semantically opaque name make sense.

Geographical names that are examples of the diversity of the South African nation are arranged under different headings indicating different linguistic and cultural groups in diachronic order starting with the Bushman and Khoe followed by names from other languages, anthroponymic names from all South African languages, names referring to shared hardship and joy and followed by names illustrating activities associated with being South African. This approach leaves no room for what W. P. Esterhuyse terms as 'cultural chauvinism' that cripples social cohesion and nation building.[7] However, cultural affinity and identity could also be building blocks, stitches, threads that keep the tapestry together as a whole providing

[5]Peter Raper, *New Dictionary of Southern African Place Names* (Johannesburg: Jonathan Ball Publishers, 2004).

[6]Heidegger in Andrea Hurst, 'Humanising Research: The Cares that Drive Researchers', *Acta Academica*, 40, 3 (2008), pp. 1–34 (p. 7).

[7]W. P. Esterhuyse, *Apartheid Must Die* (Cape Town: Tafelberg, 1981), p. 60.

there is, at a deeper level, a common goal of national unity. Where the politics of identity tend to divide and isolate people into 'theirs' and 'ours', the toponymic landscape of the country gives a non-manipulative and objective picture of who the South Africans are at a certain point in time.

Geographical names speak of human involvement through times of adversity as well as prosperity on this southern tip of the African continent. To appreciate the immense role played by all the peoples of this land, dating back to the indigenous hunter-gatherers, the Bushman; the cattle owners, the Khoe; the other indigenous African tribes and the descendants of navigators and settlers from the European continent – the semantic content of a selection of names given by these groups will serve as examples of cultural assets reflecting their life style and experience.

11.3 Toponymic Landscape

A meaningful discussion of being South African requires more than merely understanding the way things are at present. It also requires an understanding of how things were in the past and "a sense of their future, inscribed in the potential for re-interpretation (for projecting a new interpretation) which explains, enriches or expands it, allowing it to become more than what it is at present."[8] This is necessary as the policy of name changes executed by the Department of Arts and Culture through their advisory body, the South African Geographical Names Council, stems from the practice of previous governments and authorities to supplant indigenous names with what is termed colonial and apartheid names.[9] The same driving force of the politics of power and control is at work presently leading to redressing the past by using terms like 'change' and 'transformation' of the country's toponymic landscape. Naming new structures of metropolitan areas, districts and municipalities after 1994 illustrate true transformation of the toponymic landscape of South Africa. Other names in need of change are harder to transform as they have associative meanings that are part of history – a history that is, according to Said "intertwined and interdependent, and above all (carries) overlapping streams of historical experience"[10] shared by all South Africans, irrespective of race, gender, religion, age or economic status.

The matter of changing, adapting, reconstructing and standardising names started hundreds of years ago when names of Bushman and Khoe origin were phonologically adapted to other indigenous languages such as isiXhosa, isiZulu, Setswana and Sesotho, and languages like English, Dutch and German used by European settlers. Adaptations of Dutch place names to Afrikaans occurred soon after the Afrikaans

[8]Heidegger in Hurst, 'Humanising Research', p. 7.

[9]Somadoda Fikeni, 'What's in a Name? Naming, National Identity and Social Cohesion in South Africa' (Pretoria: Department of Arts and Culture, 2011), p. 10.

[10]Said, *Culture and Imperialism*, p. 378.

language became the preferred medium of communication amongst the descendants of the earlier settlers and their household staff and slaves. The Afrikaans speaking communities, today the second largest language group in South Africa, have a variety of speakers, most of whom have little in common with the Dutch (who arrived at the Cape in the seventeenth century), which is why it is often heard that members of this language group consider themselves African, having no links to Europe except maybe an inherited surname. The rationale is also that 'Afrikaans' as a language developed on this continent and is derived from 'Africa.' The toponymic landscape of South Africa reflects both the history of the indigenous population and those nationalities that came from other parts of the world either for humanistic reasons or with imperialistic motives.

11.4 The People of South Africa

It is fair to ask what South Africans have in common that could enhance social cohesion to the extent of agreeing on the naming of a place and attain the ideal of a transformed society. Geographical names describing the natural resources and topographical splendour of South Africa is a topic on its own; yet these are names appreciated by South Africans who call themselves lovers of nature, but this would also include experiences of natural disasters and wildlife. Being South African can also be seen in names that reflect attitudes, feelings, values, fears, gratitude, happiness, hopes and dreams in the light of socio-political, historical and cultural experiences in times of conflict or unity. A plea was made by Somadoda Fikeni to acknowledge the pain caused by fellow South Africans, not defensively but honestly, that diversity be accepted but that a common vision be developed by way of "a name review approach and methodology that will take into account our diverse heritage and unearthing as well as affirming the names that have been misunderstood or erased."[11]

11.5 Categories and Tendencies

11.5.1 Group A – The Bushman and Khoe

The first inhabitants of Southern Africa were the Bushman, groups of hunter-gatherers whose presence can be traced all over the country in names also given to places by other people who came into contact with them, for instance *Senwabarwana* ('where the Bushman drink'), a Municipal Village in Limpopo Province. Descendants of the Bushman have indicated that they are proud to be called

[11]Fikeni, 'What's in a Name?', p. 10.

Bushman, preferring it to the term 'San' meaning 'dog'/'thief' in the Khoe language. They do not find 'Bushman'/'Boesman' an abusive term as some politicians would like to believe. Place names like *Boesmansriviermond*, a village at the mouth of the Boesmans River, remind us of the first people who trod the South African soil. Many names adapted from their language still exist in either an adapted or translated form.[12]

When the Khoe arrived with their cattle, many conflicts took place as the newcomers and their cattle encroached on what the Bushman believed to be their territory. Eventually a new society emerged and new names were given, ultimately adapted or translated by the Dutch and English after their encounters with them because the clicks and suction sounds in the languages of the Bushman and Khoe were difficult to master. These sounds are still to be found in the names of many places as well as in languages like isiZulu and isiXhosa.[13] Around the Cape the Khoe ultimately became slaves, cooks and herders for the settlers during the eighteenth century and tried to speak the languages of their masters. The Goniqua tribe of the Khoe integrated with a Xhosa tribe during the eighteenth century and lost their racial distinction. The geographical names from the Bushman and Khoe languages are unique to this country and continent and should be treasured as heritage. The descriptive names not only reflect how the Bushman and Khoe identified places, for instance *Abbaheigaskop* ('hill where many red trees grow') and *Kraggakamma* ('place of fresh water'). They also localised and described different tribes and people, as in the village name *Gouritsmond* (at the mouth of the Gourits River, named after the Gouriquas tribe whose name means 'cattle people'). Acknowledgement of this minority group in the South African population can also be found in names given to new structures as formed through the transformation of the geographical demarcation of the country into Provinces, District Municipalities, Local Municipalities and Metropolitan Municipalities after 1994. Examples include: *!Ka !Garib Local Municipality* (Khoe 'big great river'); *!Kheis Locality Municipality* (Khoe 'a place where you live'); and *Khara Hais Local Municipality* (Khoe 'big tree'). Names from other South African languages featuring in these structures will be mentioned further on.

11.5.2 Group B – The Portuguese, Dutch, French, British and Other Groups

In Group B we have the names given by navigators and dignitaries who came ashore since the sixteenth century. Many South Africans also descend from survivors of

[12]See Peter Raper, 'Khoisan Influence on Zulu Place-Names', *Nomina Africana*, 22, 1–2 (2008), pp. 127–168. The words on the country's coat of arms are also !Xam meaning 'unity in diversity.'

[13]Peter Raper, 'Khoisan Influence Discernible in the Transformation of Place Names in South Africa', *Acta Academica*, 1, supplementation (2007), pp. 110–138 (pp. 118–121).

shipwrecks who became part of the existing society and stayed along the coast. Most of these names were given in the sixteenth and seventeenth century and can be traced in the names of businesses or in adapted or translated forms. Unlike other countries with imperialistic motives, Portuguese names in South Africa do not seem to be repetitions of similar place names in Portugal, mainly because they were navigators on their way to the East and never settled there with nostalgic longings for their home country. These names, however, honour their navigators and saints.[14] A more substantial impact than the Portuguese was made by the Dutch who came as officials of the Dutch East India Company. Although some of the Dutch names have been changed to Afrikaans, many still exist in their Dutch form as well as names honouring historical figures such as Jan van Riebeeck and Simon van der Stel; or towns such as *Riebeeckstad, Riebeeck-Kasteel, Stellenbosch, Stellenberg, Bonteheuvel, Paarl* and *Delft.*

The common trait in man to use place names from his home country to make him feel more at home in a strange country, is reflected in the names such as *East London, Newcastle* and *Belfast,* of British origin; *Amsterdam, Amersfoort* and *Delft* from the Dutch – some of which have been changed to African names in recent times; and *Bochum, Hamburg, Berlin* and *Hanover,* as German names. Many South African surnames are variations of French surnames, descending from the French Huguenots who fled from their continent for religious reasons. Their knowledge of the wine making industry led to South Africa being a leader in this sector. Many toponyms of French origin reflect this influence, mostly found in the Western Cape where the vineyards abound, for instance: *Franschoek, La Rochelle, Artois, Bordeaux* and *Marseilles.* Scottish towns like *Aberdeen, Rosebank* and *Rosetta* point to a Scottish presence in South Africa. Similarly, towns and suburbs with names like *Valhalla, Aberdeen, Edenburg* and *Brooklyn* indicate a Scandinavian, Scottish and English influence.

11.5.3 Group C – Anthroponymic Toponyms

South Africa has many anthroponymic toponyms. Giving these names that honour people is a typical trait of any group in power. It started with the Portuguese, who honoured their saints with related place names. Names of cities and towns such as *Pietermaritzburg, Dewetsdorp, Bothaville* and *Delareyville* refer to historical figures of Afrikaans origin. Not all names represent political heroes but rather historical figures with a social impact on the population. The Griquas honour their leader Adam Kok in the name *Kokstad* and the descendants of the indigenous Khoi and Bushmen honour heroes and heroines like *Sarah Baartman* and *Dawid*

[14]Barbara Meiring, 'The Stormy Seas around the Fairest Cape', *Nomina Africana*, 19, 1 (2005), pp. 1–30.

Stuurman in urban features like buildings and monuments. Toponyms honouring British dignitaries are, for instance, *Wellington, Darling, Malmesbury, Cradock, Grahamstown* and *Carnarvon*. Such a tendency is also unmistakably obvious in the names of forts bearing the names of leaders, commanders and chiefs, amounting to almost sixty while forts named after towns and settlements amount to about eight. Some forts have been declared national monuments while others do not have any status. Examples of settlements and towns built on original sites where forts were erected are: *Fort Beaufort, Fort Cox* (settlement and agricultural college), *Fort Hare* (the well known University of Fort Hare was built on this site), *Fort Jackson* (industrial site), and *Fort Nottingham* (village).[15] Fikeni reminds us that "the 100 year war of resistance was fought in this region hence the place is punctuated with so many forts."[16]

The policy stipulated in the *Handbook on Geographical Names* recommended by the South African Geographical Names Council is that names of living persons should generally be avoided in the naming of places.[17] This recommendation is not respected in the case of new names given to features as can be seen in the names of District, Local and Metropolitan Municipalities from the data lists of the Department of Arts and Culture honouring heroes such as *Nelson Mandela, Albert Luthuli, Gert Sibande, Alfred Nzo, Chris Hani, Dr. JS Moroka* and *Govan Mbeki,* among others – as part of the transformational policy bearing mostly the names of African and other indigenous leaders.

Anthroponymic place names with either a positive or negative connotation depending on the point of view, include names of clans, leaders, chiefs, commanders, generals and warriors from different cultures.[18] The interaction of culture, politics and economy is obvious when the names of prominent towns and cities are changed to honour dignitaries as in *Mokopane* (Potgietersrus) and *Tswane Municipality* (name proposed for Pretoria, administrative capital of the country that lies in the area governed by the Municipalty). A selection of urban settlements illustrates the point of a shared history of violence as most cultural groups are involved in the motivations for the names: *KwaMatiwane* (isiXhosa for 'place of Matiwane' referring to a Xhosa Chief whose warriors were destroyed by British troops in 1828, was pursued by King Shaka and executed); *Delareyville* (General De La Rey of the Boer forces during the Anglo-Boer War, shot by the police for failing to stop at a road block); *Grahamstown* (Colonel John Graham who fought against the Xhosa who suffered devastating losses. This was the military headquarters and

[15]Barbara Meiring, 'Aspects of Violence in South African Place Names', *Werkwinkel: Journal of Low Countries and South African Studies*, 5, 2 (2010), pp. 95–112.

[16]Somadoda Fikeni, 'Translate the Rich Heritage of EC', *Eastern Cape Today*, 7–13 October 2011.

[17]South African Geographical Names Council, *Handbook on Geographical Names* (Pretoria: Department of Arts and Culture, 2002), p. 5.

[18]Meiring, 'Aspects of Violence.'

Fig. 11.1 Photograph of a signpost taken on 12 September 2011 in which the name 'Makhado' had just changed back to 'Louis Trichardt' (Photo courtesy of Tia Mysoa)

centre of the 1820 British settlers); and *Makhado* (for both settlement and district, after a powerful but controversial and uncooperative Venda Chief, Makhado, who ruled in 1864).[19]

The measure of negative associative meaning that the two names *Makhado* and *Louis Trichardt* have amongst different communities in the urban areas concerned is clear with the proposal that the town *Louis Trichardt*'s name be changed to *Makhado* and the number of objections against this proposal with an eventual court case to rule whether sufficient consultation had been done in the community (Fig. 11.1). However, not all changes are met with antagonism and negative attitudes resulting in court cases, which indicate the measure of goodwill amongst South Africans with different cultures.

With regard to gender it is noticeable that women were not commemorated to the extent that men were in naming places. This is a universal trend and mainly due to the different social roles attributed to men and women. Men went to war, women stayed at home or did social work. Although women as royalty and those like the wives of commissioners and leaders were commemorated, the names of women do not appear in place names to the extent that the names of men do. This does not however reflect that women in South Africa are not liberated today. It merely reflects a social reality that women did not play the political and social roles that South African men did in the past. There are places that honour royalty (*Victoria West, Victoria Bay, Adelaide, Alice*); the wives of dignitaries during the imperialistic era (*Lady Frere, Lady Grey, Ladysmith, Ladybrand)*; the wives of commissioners,

[19]Raper, *New Dictionary*.

donators, etc. (*Cornelia, Carolina, Sannieshof, Port Elizabeth, Creighton*); and a few who refer to community leaders (*Cetuville, Masangwanaville, Mashashane, Mahloseni*).

Another group reflected in some older place names refer to pioneers who contributed to the development of the country as pathfinders, frontiersmen and entrepreneurs who paved the way for others to follow. An example is *Jabavu*, suburb in Soweto named after Davidson Don Tengu Jabavu – co-founder of the ANC, first black academic in the country, professor in Latin and African Languages, editor of the first full-fledged newspaper for the black community, and son of the Cape rebel John Ntengu Jabavu who was intimidated and persecuted because he refused to support the expanding of the British empire. He was honoured at the Anglo-Boer Monument in Paarl where any inhabitant of the Cape who stood up for democratic principles of freedom and justice against the British imperialistic war of 1899–1902 is honoured. He was also instrumental in the foundation of the University of Fort Hare. The Jabavu name was also carried on by Noni Jabavu, grandchild of Davidson Jabavu, by being the first Xhosa woman who published biographies about her family.[20]

11.5.4 Group D – Names of New Entities

The creation of new entities in the form of municipal/district/metropolitan areas and freedom/trans frontier parks led to innovative naming to reflect the new South Africa. A few of the new names were mentioned under the discussion of anthroponymic names and those with a Bushman and Khoe heritage. Other names include names referring to natural features of the fauna and flora of the areas. As an example names from the Eastern Cape Province will be used as presented by the Department of Arts and Culture (undated Data lists). Of the six District municipalities, three are anthroponymic and commemorating struggle heroes (*Chris Hani, O.R. Tambo* and *Alfred Nzo District Municipalities)*. The other District Municipalities have names like *Ukhahlamba* (isiZulu name for the Drakensberg referring to 'pointed spears') or *Cacadu* (name of a river, Khoe 'bull rush river') (Fig. 11.2).[21] The *Nelson Mandela Bay Metropolitan Municipality* covers Port Elizabeth and surrounding towns. The names of Local Municipalities vary from commemorative names to descriptive names of which the majority are indigenous. The District Municipality with the most balanced linguistic naming practice is *Cacadu* where the names of Local Municipalities are in Khoe and in two of the other three official languages of the province, that is, isiXhosa (close to Khoe),

[20]Noni Jabavu, *Drawn in Colour: African Contrasts* (London : John Murray, 1960), p. 14;
Martiens van Bart, 'Dié rebel ook vereer by ABO-Monument op Laborie', *Nuusbrief: SA Akademie vir Wetenskap en Kuns*, September 2005.

[21]For other explanations see Raper, *New Dictionary*, pp. 9, 76. The latter name is presently under scrutiny due to possible confusion with similar names for features in KwaZulu-Natal.

Fig. 11.2 Welcome signpost to Cacadu District Municipality, situated in the Western Portion of the Eastern Cape Province, South Africa (Photo courtesy of Eco-Tec company, Durban)

English and Afrikaans. Examples of isiXhosa names are *Ndlambe* (name of a king), *Makana* (name of the prophet/leader in the attack on Grahamstown in the nineteenth century) and *Ikwezi* ('a star'), while names derived from Khoe are *Camdeboo* ('green hollow') and *Kouga* (name of a river 'many hippos'). English names are *Blue Crane Route* and *Sundays River Valley* while *Baviaans* is the Dutch for 'baboon' and could not count as an Afrikaans name.

The names of other Local Municipalities in the Eastern Cape collectively are twenty four in isiXhosa, three from Khoe, two in English and none in Afrikaans. The Local Municipalities in the *Chris Hani District Municipality* and the *Alfred Nzo District Municipality* are all in isiXhosa, indicating successful transformation of the naming of Eastern Cape municipal structures, yet a violation of the principle of the equity of languages according to the constitution of the country. However, the majority of the population and language groups in an area should also be taken into account. Similar patterns of naming municipal structures in other provinces reflect the transformational drive that ultimately changed the toponymic landscape of South Africa. This correlates with the changing profile of the South African society in its transformational process.

The new Trans Frontier Parks (also referred to as 'freedom parks') is a cross border endeavour correctly described by Nelson Mandela as: "I know of no political movement, no philosophy, no ideology, which does not agree with the peace parks concept as we see it going into fruition today. It is a concept that can be embraced by all."[22] Needless to say that the names of these parks bear the key to reconciliation

[22]Cited from http://www.peaceparks.org/(visited 4 April 2014).

and transformation because of the wide spectrum of associative meaning in the combined names. The conservation areas shared by the southern countries like South Africa, Swaziland, Mozambique, Botswana, Namibia, Zambia and Lesotho lead to cross border initiatives to establish conservation areas for the protection of southern African game and overall biodiversity. This not only led to new indigenous names given but ensured economic stability for the local communities due to employment opportunities in the erection of the infrastructure and major opportunities for the development of tourism. This is an example of transformation at its best, a fresh new approach to conservation with names for these seven parks from most of the indigenous languages. Among these toponyms are: *Kgalagadi* (Setswana 'semi-desert', including parts of Namibia, Botswana and South Africa); *Greater Mapungubwe* (named after and including a famous archaeological site in the Limpopo Province, part of Zimbabwe and Mozambique, Tshivenda 'place of the jackals'); and *Great Limpopo* (named after the river between Zimbabwe and South Africa, Ndebele 'rushing water', covers areas of South Africa, Mozambique and Zimbabwe).

11.5.5 Group E – Names Associated with Experiences of Hardship or Happiness

Similarly to the former sub-heading these descriptive toponyms laden with historical significance refer either to violence or goodwill implicating a human quality of man as a being who can be violent but also strives for goodwill and peace, a bipolar situation involving victims and executioners cutting through all social groups.[23] A name like the historical town and residential area *Bethulie* is exemplary. Formerly called *Moordenaarspoort* (Afrikaans for 'murderers' gate'), it is an historical name for a place where many Khoe from the Griqua tribe and Bushman were murdered by the Basotho, later a mission station, then a town, now *Bethulie*, ironically meaning 'a virgin' in Hebrew, given by the French Missionary Society. The place is also associated with the graves of women and children who died in the concentration camp during the Anglo-Boer war. This background description makes the name both descriptive and connotative, allowing further interpretation regarding human involvement. The resilience and strength of South Africans was also built through survival of extremes in weather and natural dangers encountered as reflected in many descriptive toponyms for natural features.[24] For example, the urban area of *Mabopane* means, in Northern Sotho 'coldness of the veld', 'despair', 'slow death' and 'make bricks by hand.'

[23]Barbara Meiring, 'Discourse Patterns at Social Cohesion Campaigns', *Language Matters*, 41, 1 (2010), pp. 41–57 (p. 47).

[24]Raper, *New Dictionary*; Meiring, 'Aspects of Violence', pp. 100–110.

Associative meanings attached to names often add to the semantic content of the name. Many names do not have any lexical reference to violence but their descriptive backing is associated with war or other forms of violence, for instance social injustice leads to protest actions in urban areas and many of these names are associated with the presence of police action or riots as well as clashes between political factions leading to hostile actions. Examples where the lexical meanings of names express the opposite of the associative meanings, are townships like *Boipatong,* meaning 'place of shelter', associated with an attack by the Inkatha Freedom Party with fatal results and trauma for young children which would shape their lives, both negatively and positively[25]; *Sharpeville, Sebokeng* ('place of gathering' in Sotho) and *Langa* ('sun' in isiXhosa) are associated with protests resulting in violent clashes with the police with fatalities and many hurt.[26] Such examples indicate how these names reflect a shared history and heritage that shaped a collective experience of what it means to be a South African.

In spite of the hardships reflected in the above mentioned place names, being South African is also characterised by messages of peace, hope, safety, well-being and happiness as reflected in names of urban settlements like: *Agisanang* (Setswana for 'live together in peace'); *Kagiso* (Setswana for 'place of peace'); *Kgotsong* (Southern Sotho for 'place of peace'); *Khanyiso* (isiXhosa for 'peace'); *Vrede* (Afrikaans for 'peace'); *Giyani* (Shangaan for 'place where people dance for joy'); *Guguletu* (isiXhosa for 'our pride'); and *Duduza* (isiZulu for 'place of comfort/consolation').

11.5.6 Group F – Cultural and Economic Activities

Names reflecting differences in language, culture and religion are indicators of the differentiated South African society, however, economic activities indicate how South Africans explore possibilities and opportunities that strengthen the nation as a whole.

11.5.6.1 Languages

South Africa has eleven official languages, nine of African origin, Afrikaans and English. Indian languages are also spoken. Efforts to learn and teach some of the nearly extinct Khoe and Bushman languages are made by the descendants of these people. With so many languages it is inevitable that there would be names for different places that refer to the same thing, for instance 'place of the leopard/cheetah' is found in *Mangaung* (Southern Sotho), *Mahloseni* (Xitsonga),

[25]Pierre de Villiers, 'Captain Courageous', *Indwe*, June 2010.

[26]Raper, *New Dictionary*, p. 27.

Mankweng (Northern Sotho), *Luipaardsvlei* ('marsh of the leopard'), *Tygervallei* and *Tierpoort* – the names with 'tier' (tiger) actually refer to the leopard or cheetah. We also find 'place of red ochre/red ground/red sand' in *Ntsumaneni* (Shangaan), *Luvhundini* (Tshivenda), *Awakhaes* (Khoe), *Roodezand* (Dutch), *Bomvini* (isiZulu), *Rooilaagte* (Afrikaans) and *Red Hill.*[27] Similar varieties exist with other words from the fauna and flora and natural of the country being the element of description.

11.5.6.2 Economic Activities

South Africans are well known for their economic abilities and especially their agricultural expertise. Taking only cattle farming as an example we find names dating back to the Khoi cattle owners who provided the navigators with fresh meat, for instance names referring to 'the cattle people', the Gouriquas who lived in what is today *Gouritsmond*, a village named after a river that was named *Rio das Vaccas* ('river of cattle') and *Rio dos Vacqueros* ('river of the cattle people') by the Portuguese navigators. Other examples are *Vleesbaai* ('bay of meat'), also named *Angra das Vaccas;* and *Patensie,* a phonological adaptation of Bushman 'where the cattle stays'. Economic activities are also reflected in names referring to the processing of indigenous fruits, for instance the marula and the ripe little black berries of another wild plant (isiXhosa *msobo*, 'deadly nightshade'), used for jam making which is unique to South Africa. The little place, *Msobo,* in the Eastern Cape Province gets its name from this plant. The marula is a national fruit associated with the country and used in liqueurs, jams and other products for export purposes. This industry is a source of income for rural women who use the fruit for making jelly, syrup, beer and dried rolls. The nuts from the pith are said to be high in healthy oils. The fruit presence of the marula is reflected in names like *Nhlowa* (Shangaan for marula juice, the unfermented, undiluted form of marula liquor), a tributary of the *Sabie River; Maroelakop* ('marula hill') and *Maroelapan* ('marula dam'). Other places are *Ceres*, named after a Roman goddess of agriculture, a place known for its fruit and juice industry; *Citrusdal*, where many of South Africa's export oranges come from; *Golden Valley*, named after the colour of the ripe fruit, mainly apricots, grown there under irrigation from the Great Fish River.

11.5.6.3 The Mining Industry

South Africa is well-known for its mining activities, started many decades ago and was the root of many wars between South Africans and countries with imperialistic motives. To many South Africans the gold mining industry meant a source of income and therefore shaped their lives, often having to spend months at a time away from their homes where their wives and children had to cope without them. For others

[27] Raper, *New Dictionary*.

it meant wealth and prosperity and a lifestyle to go with it. *Thabazimbi* (Setswana for 'mountain of iron') is a mining town established in 1953, yet remains of clay pots and the melting of iron to be found on the surrounding hills are proofs of mining activities dating back to as far as 1400–1600 long before the town was there or had a name; *eMalahleni* (the new name of Witbank) meaning 'a place of coal' in isiZulu, isiXhosa and isiNdebele. In a name like *Gauteng* (Sesotho for 'place of gold' the phonological similarity to Afrikaans 'goud' is clear in the Sesotho word); *Egoli* (isiZulu for 'place of gold'); *Gold Reef City, Goudvlakte, Pilgrims Rest* and *Barberton* are associated with the mining of gold. *Abbots Hill* was named after Abbot, a prospector in the Barberton gold industry in the nineteenth century; *Allanridge* after prospector Allan Roberts; and *Barberton* after a prospector, Graham Barber. The diamond industry is associated with *Kimberley* and *Cullinan*. Interestingly enough the African name for Kimberley is *Teamane*, a phonologisation of Afrikaans 'diamant' which means 'diamond.' Copper is/was mined at *Musina* (meaning 'copper' in Tshivenda) and *Copperton*. Names reflecting other mining activities can also be identified.

11.5.6.4 Religion

South Africans are often described as a nation with a deep-rooted religious heritage consisting of various forms of worship. The role of missionaries from various countries in the planting of mission stations and the influence it had on the lives of people cannot be evaluated or measured except by word of mouth from those and their descendants who were subjected to it, voluntarily or not. One point of view is that "the good intentions of European colonialists and missionaries [in South Africa] became awkwardly entangled with exploitation, separatism and supremacy."[28] Others believe in the sincerity of missionaries who came to spread the gospel to save souls and uplift the people. Apart from well-known Biblical names in the South African urban toponymic database noted by Naftali Kadmon such as *Bethlehem, Samaria, Bethel, Eden, Hebron, Berea, Bethanie, Bethulie, Kanaan, Libanon, Mara, Morija, Nazareth, Pniel* and *Salem*,[29] there are also many mission stations, for instance *Pella* (Roman Catholic mission station); *Adams Mission* (established in 1836, destroyed by Dingane, rebuilt in 1839, an educational institution named after Dr. Newton Adam, American missionary); and *Mariannhill* (founded in 1882 as a Trappist monastery, now a Roman Catholic mission station).[30]

[28]Gerald L'Ange, *The White Africans: From Colonisation to Liberation* (Johannesburg : Jonathan Ball, 2005), p. 173.

[29]Naftali Kadmon, 'Research in Toponymy: Biblical and other Hebrew Place Names in Southern Africa', in P.S. Hattingh, Naftali Kadmon, Peter E. Raper and I. Booysen (eds), *United Nations Group of Experts on Geographical Names Training Course in Toponymy for Southern Africa* (Pretoria: Department of Geography, University of Pretoria, 1993), n.p.

[30]Raper, *New Dictionary*, p. 253.

11.6 Transformation and Change

Whereas most names for the previously mentioned new entities/structures are African names (not necessarily the names of persons, with a few Khoe, English and Afrikaans names), the process of transformation has been successful in representing the majority of South Africans in these names. The country has a colonial past and a liberation struggle still fresh in the memory of many South Africans. Efforts to bring about reconciliation is a strategy to reach an outcome of transformation so that every South African can share in the joy of being a liberated nation, free from divisional structures between different groups of people. Geographical names are targeted to be part of the liberation struggle and establishment of an African identity. Similar political moves were made by previous authorities but honouring the legacy of the past as a gesture of reconciliation is not always a good reason to justify the status quo, especially if some of the older names are abusive to a certain community. Names like *Verwoerdburg* had to be changed due to the connotation of Hendrik Verwoerd as the leader of the policy of apartheid and all its abusiveness. As part of the history of South Africa, the removal of such names is a strategy to develop a new order (transformation), a kind of evolution from one social order to a new more just society. To maintain the delicate balance between a goal of national unity and providing for social diversity demands that the politics of identity should not be a precondition for the strategy to reform and renew. Successful transformation of society, ultimately having social cohesion as the foundation of a South African society implies a certain social consciousness. If to borrow the words of Frantz Fanon in this context, "unless national consciousness at its moment of success was somehow changed into a social consciousness, the future would hold not liberation but an extension of imperialism."[31]

This implies that a consciousness of social and political need should underpin recommendations to change a specific place name. The question is how South Africans see the toponymic landscape of their country – is it a reflection of all its peoples or only the majority in power? Are older names abusive in the sense that they have negative associative meanings of being part of a previous era or does the abusiveness lie in the lexical meaning of the word? Reactions to recent changes of geographical names, either by way of replacement with older established names, the translation of a name or innovation of new names, were met with mixed attitudes from communities. Strong opposition came from communities who felt they were not consulted sufficiently while others welcomed the new name. Objections will always be unavoidable as South African society is not homogenous on any level. Until there is, in spite of a cultural identity, a national strategy aimed at attaining national unity, filtering through to all communities, promoting reconciliation through tolerance, a change of attitudes based on trust and goodwill, the process of change cannot be termed 'transformation.' Reactions to

[31]Fanon was quoted in Said, *Culture and Imperialism*, pp. 323–325.

standardisation, that is, orthographic adjustments to names with incorrect spelling, are, however, minimal. The dynamics behind the changing of a country's names is discussed in more detail in Jenkins (2007), Coetser (2006) and Meiring (1996).[32]

11.7 Conclusion

The geographical names used as examples are no more than indications, reflections of the South African way of putting their experiences, activities, feelings, observations of their surroundings into words – words that are names. Being South African is being connected, being part of the historical experience reflected by the toponymical tapestry of the land. Many of the names reflect more about the beauty or harshness of the mountains, sunsets, coastline, pastures, plains, fauna and flora of the country than about the people. Yet, many of these features are chosen by South Africans for street names in urban settlements as themes for suburbs. Unlike other conventional uses of language, names do not lie, not even when they appear semantically opaque like names derived, translated or adapted from Bushman and Khoe names.

The essential element in this article is man, who is the name giver, the inhabitant of a place and the creator of his personal space within a natural and cultural environment. Man, as a governing authority, is responsible for the demarcation of his living areas eventually constituting separate provinces, states, municipalities, recreational, industrial and housing areas, however, none of this is ever constant, as life, which includes man and his mobility as a cultural being, is a dynamic force being susceptible to change. This is reflected also by language shifts, political attitudes, cultural diversity and social change that influence the choice or rejection of a toponym. In the words of the Ghanaian philosopher Kwasi Wiredu, describing the essence of being (South) African as "[t]he sense of solidarity and fellowship which, as it were, spills over from the extended family to the larger community and the well-known spontaneity of our people [...] combine to infuse our life with a pervasive humanity and fullness of life which visitors to our land have always been quick to remark."[33]

[32]Elwyn Jenkins, *Falling into Place* (Cape Town: David Philip, 2007); Attie Coetser, 'Africa's Onomastic Reaction to Colonialism: From Countries to Municipalities', *Nomina Africana*, 20, 1–2 (2006), pp. 16–41; Barbara Meiring, 'New Names for a New South Africa', in W. F. H. Nicolaisen (ed.), *Proceedings of the XIXth International Congress of Onomastic Sciences* (Aberdeen: the University of Aberdeen, August 4–11, 1996), pp. 241–247.

[33]Kwasi Wiredu, *Philosophy and an African Culture* (Cambridge: Cambridge University Press, 1980), p. 21.

Chapter 12
M(g)r. De Hemptin(n)e, I Presume? Transforming Local Memory Through Toponymy in Colonial/Post-Colonial Lubumbashi, DR Congo

Johan Lagae, Sofie Boonen, and Donatien Dia Mwembu Dibwe

Abstract In 2010, the Congolese city of Lubumbashi celebrated the centennial of its foundation. Numerous activities were organised on the occasion, most of which celebrated rather than critically assessed the city's colonial past. It was precisely during this year that Moïse Katumbi Chapwe, the current provincial governor of Katanga and one of Congo's rising political stars, decided to re-baptise one of the central avenues in Lubumbashi's city centre, changing its name from *Avenue de Tabora* to *Avenue Mgr. Jean-Felix de Hemptinne*. Making a tribute to the head of the missionary congregation of the Benedictine fathers and a notorious figure of Lubumbashi's colonial history was a rather remarkable decision, all the more so since this particular avenue had been the symbolic axis of Belgian colonial power during colonial times. One can wonder then to what extent Katumbi's initiative was an act similar to what the prominent Africanist scholar Valentin Mudimbe, who was born and trained in Lubumbashi, described in his 1994 book *Les corps glorieux des mots et des êtres* as the conscious transformation of (local) memory through a new colonial toponymic inscription in space.

Drawing on archival research, fieldwork and local history, we will demonstrate in this chapter that the urban history of Lubumbashi offers a number of telling examples of how power has been imprinted in the minds of the city's inhabitants via the urban text that is constituted by names of streets and avenues. Focusing on a limited number of examples, the chapter will illustrate that such practice was, however, not exclusive to the colonial, but continued into the post-independence era. As such, we will argue, the changing toponymy of Lubumbashi offers us a particular insight in the shifting position of the city and, by extension, the province

J. Lagae (✉) • S. Boonen
Department of Architecture and Urban Planning, Ghent University, Ghent, Belgium
e-mail: johan.lagae@ugent.be; sofie.boonen@ugent.be

D.D.M. Dibwe
History Department, University of Lubumbashi, Lubumbashi, Democratic Republic of the Congo
e-mail: dibwe.mwembu@unilu.ac.cd; donatiendibwe@gmail.com

L. Bigon (eds.), *Place Names in Africa*, DOI 10.1007/978-3-319-32485-2_12

of Katanga, in the minds of local as well as central authorities in Congo. Resituating Katumbi's initiative of renaming the *Avenue de Tabora* in this larger narrative will allow us to define it as an impulsive rather than a thoughtful act, more informed by post-colonial amnesia than by a profound engagement with colonial urban memory.

Keywords Lubumbashi • (Post-)colonial toponymy • Colonial urban memory • Belgian Congo • Katanga • Tabora • Colonial urban Planning

12.1 Introduction

In 2010, the Congolese city of Lubumbashi celebrated the centennial of its foundation. Numerous activities were organised on the occasion, most of which celebrated rather than critically assessed the city's colonial past.[1] It was precisely during this year that Moïse Katumbi Chapwe, the current provincial governor of Katanga and one of Congo's rising political stars, decided to re-baptise one of the central avenues in Lubumbashi's city centre, changing its name from *Avenue de Tabora* to *Avenue Mgr. Jean-Felix de Hemptinne*. This decision constitutes a remarkable gesture, as the previous name of the avenue had helped define this particular thoroughfare as the symbolic axis of Belgian colonial power during colonial times. How are we to read Katumbi's initiative, which brings to the fore the head of the missionary congregation of the Benedictine fathers who was a notorious figure in the city's colonial history?

To answer this question, we will discuss a number of telling examples of how power has been imprinted over time in the minds of the *Lushois*, the city's inhabitants, via the urban text that is constituted by names of public places, streets and avenues.[2] We will show that such practice continued from the colonial into the post-independence era, and that it most often was a deliberate act reflecting the shifting position of the city of Lubumbashi and, by extension, the province of Katanga, in the minds of local as well as central authorities in Congo. Even if new names can be

[1]As part of many official celebrations, the University of Lubumbashi also organised an international conference on the city's history, the results of which were published as Maurice Amuri Mpala-Lutebele (ed.), *Lubumbashi. Cent d'ans d'histoire* [*Actes du Colloque internationale 'Lubumbashi, cent ans d'histoire. Littérature, cultures urbaines, débats intellectuels'*, 9–11 September 2010, Campus de l'Université de Lubumbashi] (Paris: L'Harmattan, 2013).

[2]In this respect, this article builds further on the preliminary study of Lubumbashi's changing toponymy, prepared by Prof. Michel Lwamba Bilonda of the University of Lubumbashi, on the basis of a large collection of documentation of the city, part of which was provided by Johan Lagae during a visit to the city in 2000. See Michel Lwamba Bilonda, *Histoire de l'onomastique d'avenues et des places publiques de la ville de Lubumbashi (de 1910 à nos jours)* (Lubumbashi: Presses Universitaires de Lubumbashi, 2001).

contested, denied or just simply ignored by the *Lushois* navigating the city daily,[3] tracing the genealogy of colonial and post-colonial toponymy helps us to understand the myriad forces that made and shaped a city like Lubumbashi. By situating Katumbi's renaming of the *Avenue de Tabora* in this larger historical narrative, we will argue that his initiative was an impulsive rather than a thoughtful act.

12.2 Installing an Urban Grid on a *terra nullius*

Although some discussion exists on the topic, most scholars today agree that the city of Lubumbashi, or Elisabethville as it was called during the colonial era, was founded on a *terra nullius*, a vacant territory. The land where general Wangermée, the then vice-governor of Katanga, decided to install a colonial outpost in 1910, did belong to certain local chiefs, but no pre-colonial settlements existed on the specific site.[4] Wangermée's choice was in part informed by scientific missions that had already indicated the promising prospects for mining in the region. Because of its immense richness of natural minerals, Katanga had already been described even as a 'truly geological scandal.'[5] In that respect, Lubumbashi provides the model *par excellence* of the colonial city as presented in colonial historiography: a place where history only started with the arrival of the white coloniser, who gave the territory its meaning and *raison d'être*.[6] Similarly, the urban form of Lubumbashi provides evidence of what was perhaps the most explicit application of the theoretical scheme of the ideal colonial city in the Belgian colony: looking at a 1927 plan,[7] one can very easily distinguish the 'European town' from the '*cité indigène*' or 'native town', the

[3]In many cases, the *Lushois* indeed continue to use former street names. This is, for instance, the case for the *Avenue du Kasaï* de-baptised in 1992 as *Avenue Munongo*, the *Avenue de la Révolution* de-baptised as *Avenue Jean Nguz a Kar I Bond* and the *Boulevard de la Katuba* de-baptised as *Avenue Kyungu wa Ku Mwanza*.

[4]The discussion is due to different views on what should be considered as 'the site' of the city of Lubumbashi. If we consider the initial construction site of Lubumbashi in 1910, than there is a general agreement that this was indeed a vacant terrain since the city was not built on top, nor constituted a transformation, of existing villages. But if one refers to the whole surface of the present day city, then it is clear that there were indeed some villages (Kimbwambwa, Tshamalale, Kabulameshi, etc.) within this territory.

[5]For a popularising account of the early decades of Katanga and Elisabethville, see René-Jules Cornet, *Terre Katangaise* (Brussels: Ed. M. Lesigne, 1950), p. 54. For a more sound, scholarly discussion, see Mpala-Lutebele, *Lubumbashi*.

[6]The idea that no cities existed in Africa in precolonial times has by now been successfully challenged, see for instance Catherine Coquery-Vidrovitch, *Histoire des villes d'Afrique noire: des origines à la colonisation* (Paris: Albin Michel, 1993). Moreover, archaeological research has made it abundantly clear that the Katanga region has a very long history of trade and inhabitation.

[7]Africa Archives, Brussels (hereafter AA), Archival funds cartography, (22)10A and (22)10B, 'Comité Spécial du Katanga. Plan d'Élisabethville et environs. Scale 1:10.000', 12 January 1927.

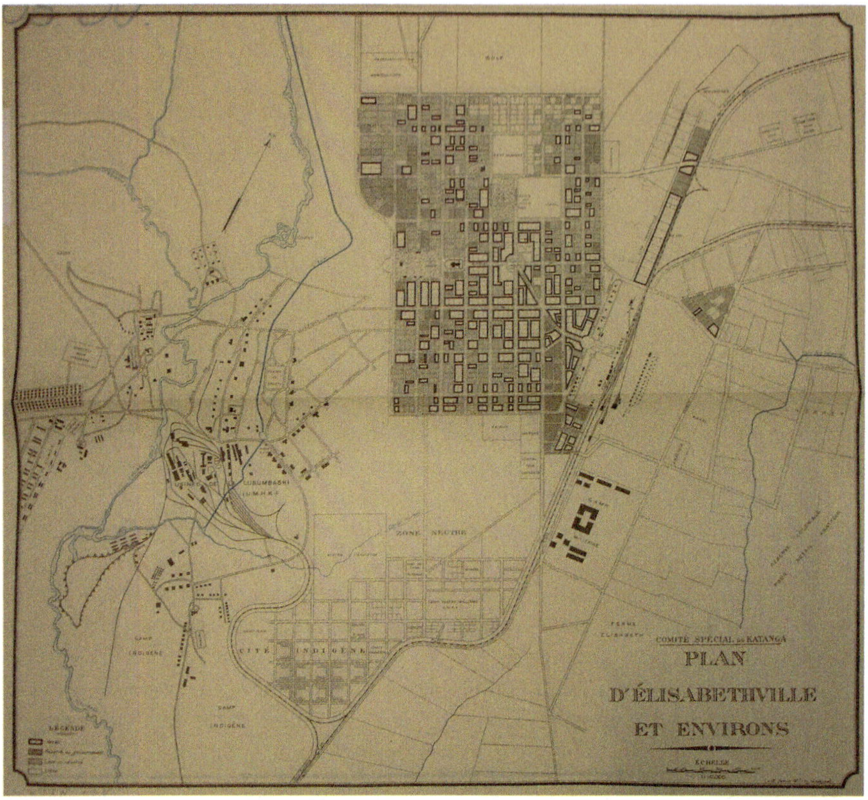

Fig. 12.1 Plan of Lubumbashi, 1927 (Archival fonds cartography, Archives Africaines, Ministry of Foreign Affairs, Brussels)

two being neatly separated by a no man's land, labelled '*zone neutre*',[8] complying with a common planning practice in sub-Saharan African colonies at the time (Fig. 12.1).[9]

Both parts of the city were laid out according to an orthogonal grid. In the European part of town, such an organisation of space facilitated the '*mise en valeur*' of the terrain. As Bruno de Meulder has argued, one can indeed view the early urbanisation of Lubumbashi as a real estate operation rather than a genuine

[8]On spatial segregation along racial lines in Lubumbashi, see Johan Lagae, Sofie Boonen and Maarten Liefooghe, 'Fissures dans le "cordon sanitaire": Architecture hospitalière et ségrégation urbaine à Lubumbashi, 1920–1960', in Mpala-Lutebele, *Lubumbashi*, pp. 247–261.

[9]For a discussion of the practice of urban segregation on a worldwide scale in these years, see Carl Nightingale, *Segregation: A Global History of Divided Cities* (Chicago: University of Chicago Press, 2012).

act of creating a pleasing urban landscape through planning and design.[10] It is no coincidence that the first articles discussing the role of architecture in the Belgian colony, which appeared in one of the leading professional journals *Tekhné*, contained harsh critiques of these spatial developments in Lubumbashi:

> They pretend to turn Elisabethville into a beautiful city and what do they do? First, they start by tracing streets and avenues along that despicable model of the American grid, that nonsensical checkered pattern [...] Then, they let things go, leaving huts in rammed earth, shacks in metal sheeting, and barracks in wooden planks spring up like mushrooms.[11]

The use of the grid as the organising model for the first native quarters has a different origin. Echoing the lay-out of labour camps, the orthogonal ordering of streets resulted here from considerations of economy, efficiency, control and hygiene, and the typology of the modest structures built to house the African population closely followed design guidelines that were current in the mining compounds of Johannesburg, a built environment with which Belgian colonial officials were well acquainted.[12] The first native town in Lubumbashi to be built after the introduction of the '*zone neutre*', today known as the *Commune Kamalondo*, was actually constructed around a labour camp of the city's mining enterprise *Union Minière du Haut-Katanga* (UMHK). The contour of this existing settlement, the so-called *Camp Robert Williams*, actually defined the dimension of the basic unit of the *cité*'s orthogonal grid.

A rare, unpublished photograph of the construction of the *Commune Kamalondo*, dating from the early 1920s, provides a particular insight on the impact on the ground of this abstract colonial ordering of space.[13] Depicting the construction of one of the avenues of the *cité*, the image reminds us of the fact that Lubumbashi's territory consists of rather arid soil, dotted with termite hills of often impressive scale. Its caption indicates that prisoners were used to clear the terrain from these natural obstacles in order to create a *tabula rasa* on which the orthogonal grid could

[10]See the chapter 'Het Comité Spécial du Katanga en de industriestad als immobiliënprobleem: Elisabethstad en Jadotstad' in: Bruno de Meulder, *Kuvuande Mbote. Een Eeuw Architectuur en Stedenbouw in Kongo* (Antwerp: Hautekiet/deSingel, 2000), pp. 71–92.

[11]V. H.-D., 'Les trentes maisons d'Elisabethville', *Tekhné*, 37 (1911), pp. 418–420 (p. 420) (our translation).

[12]In 1921, for instance, the then governor general Maurice Lippens had visited the mining compounds in Johannesburg and promoted them actively as a model to be implemented in Lubumbashi. See: Ferdinand Grévisse, *Le Centre Extra-Coutumier d'Elisabethville: Quelques aspects de la politique indigène du Haut-Katanga industriel* (Brussels: Institut Royal Colonial Belge, 1951), pp. 5–6. On the lay-out of labour camps in the Belgian colony and the influence of foreign practices in this regard, see Bruno de Meulder, *De Kampen van Kongo: Arbeid, Kapitaal en Rasverdeling in de Koloniale Planning* (Amsterdam/Antwerp: Meulenhoff/Kritak, 1996).

[13]AA, GG. 15.953 'Cité indigène Élisabethville avec plans/cartes et photos', photograph of *Commune Kamalondo* – building of a street, early 1920s.

Fig. 12.2 Photograph of Lubumbashi's *Commune Kamalondo*, building of a street, early 1920s (GG. 15.953, *Cité indigène Élisabethville avec plans/cartes et photos*), Africa Archives, Ministry of Foreign Affairs, Brussels)

easily be applied (Fig. 12.2).[14] Looking at early urban maps of Lubumbashi, one cannot but notice the peculiar contour of the *Commune Kamalondo*.[15] Close analysis reveals that it was actually defined by the particular topography of the site, as well as by the curves of the already existing railroad track linking the main line to the site of the UMHK. In other words, the abstract urban grid immediately became site-specific.

12.3 Inscribing a Colonial Lexicon on the Urban Grid

Projecting an orthogonal grid on the land was one way of introducing a disciplining colonial order on an African territory. As recent research into Lubumbashi's urban history reveals, this grid formed the physical framework for particular strategies for governing urban space.[16] In the 'European town', for instance, a functional

[14]Contemporary accounts confirm that prisoners were used as labour for public construction works. See, for instance, Henri Segaert, *Un terme au Congo belge: Notes sur la vie coloniale 1916–1918* (Brussels : A. Van Assche éditeur), p. 50.

[15]The *Commune Kamalondo* was called quarter *Albert I*[er] in colonial times.

[16]This forms part of the PhD research of Sofie Boonen at Ghent University in which she investigates Lubumbashi's urban history from a spatial perspective.

zoning was projected on the grid, distinguishing particular areas for residential purposes, others for commercial or leisure activities, while certain sites with high visibility were destined for public facilities such as a post office, a court of justice or the cathedral. Given the fact that Lubumbashi's urban population was extremely heterogeneous in the first two decades, some observers described it as a city constructed by '*des gens d'ailleurs*' ('people from elsewhere').[17] In this context, local authorities tried to control migration flows via strict selection criteria, facilitating access to land to Belgian migrants while trying to block it for non-Belgian applicants, thus trying to turn Lubumbashi into a genuine Belgian colonial town. It was a strategy that would, however, only be partly successful.[18] In accounts of the interwar years, the city is often described as a '*ville cosmopolite*', pointing at the strong presence of non-Belgian European communities such as, among others, Greeks, Italians, Portuguese and Jews.[19]

In this context, it should not come as a surprise that urban toponymy formed a tool to strengthen the Belgian identity of the urban landscape. In his 1994 autobiographical account *Les corps glorieux des mots et des êtres*, the prominent African scholar Valentin Mudimbe, who grew up in Lubumbashi, argues how the naming and renaming of the African territory in a colonial context was deeply embedded in an act of transforming (local) memories. "A colonial policy of toponymy that existed between 1885 and 1935", Mudimbe writes, was transforming African locations into "signs of monarchist allegiance." Elisabethville of course refers to the name of the Queen, wife of King Albert I. Yet, it was also "a radical reorganization of an ancient site and of its political makeup, but more importantly, generally, it indicated the invention of a new site and body whose routes and movements reflected a new political economy."[20] In this respect, Mudimbe also points our attention to how the racial segregation of urban space was reflected in the colonial urban lexicon.

[17] A. De Hertogh, 'Page coloniale: Pour commencer', *Tekhné*, 1, 1 (1911), pp. 10–11.

[18] In contrast to the rest of the Belgian Colony, where a '*colonisation de peuplement*' ('white settler colonialism') was discouraged, a propaganda program was set up to encourage a permanent settlement of Belgians in the Katanga region. Several migration files conserved in the Africa Archives reveal how Belgian colonists were favoured above non-Belgian European migrants.

[19] At the end of the 1940s approximately 30 % of the city's white population were non-Belgians. See, for instance, Noel A. Van Malleghem, 'Livre IV. L'Urbanisation d'Élisabethville', in Ministry of Colonies (ed.), *Urbanisme au Congo* (Brussels: Les Éditions de Visscher, 1950), [n.p.]. Also the African population of the city was very heterogeneous. In order to respond to the lack of local labour, large enterprises implemented a recruitment policy which resulted in migration flows with African workers coerced to come from Rhodesia and South Africa and, from the 1920s onwards, also from different regions in Congo, Rwanda or even Senegal.

[20] Valentin-Yves Mudimbe, *Les Corps Glorieux des Mots et des Choses: Esquisse d'un Jardin Africain à la Bénédictine* (Montréal/Paris: Humanitas/Présence Africaine, 1994). We will refer here to Mudimbe's other, English language book, *The Idea of Africa*, in which the part of his discussion of colonial toponymy is also included (Bloomington: Indiana University Press, 1994), p. 134.

Indeed, in Lubumbashi, the spatial division of the colonial city in separate neighbourhoods for Europeans and Africans was made explicit by naming the extreme southern avenue of the European quarter bordering the '*zone neutre*' the *Avenue Limite Sud* (today *Avenue Likasi*). Urban segregation along racial lines was thus subtly embedded in everyday parlance and informed how one navigated the city. When tracing the colonial toponymy in historical urban maps of Lubumbashi, one cannot but notice how its place- and street names, like elsewhere in colonial sub-Saharan Africa, as Bigon and Njoh has argued, were drawn from a "Eurocentric spatial and environmental design lexicon" and introduced a "sense of order into what was otherwise a nondescript space (Fig. 12.3)."[21] Apart from the name of the city itself, Elisabethville, several other names referred directly to the Belgian Royal Dynasty which had played such a crucial role in the Belgian colonial venture in Central Africa. One street as well as the hospital for Africans was named after King Leopold II, the 'founding father' of the Belgian Congo, while another avenue referred to his successor King Albert I who also gave his name to the first native town, today known as the *Commune Kamalondo*. His wife, Queen Elisabeth, also had a street, the *boulevard* on which the governor's residence was situated, named after her, as well as the hospital for Europeans as she had played an important role in setting up a programme of strengthening medical facilities in the colony during the 1920s. The name of their son, Prince Léopold, was given in the 1920s to a street linking the African to the European quarter, while his wife's name *Reine Astrid* was deliberately given in the late 1940s to the avenue that was soon to become one of the city's most prominent axes, as plans already existed to build a cultural complex and sport infrastructure at one of its end points.[22] The European quarter's main square, where a large proportion of the official buildings were situated, and the one avenue that cut diagonally through the urban grid in order to link this place to the railway station were tellingly named the *Place Royale* and the *Avenue Royale*.

Similarly, toponymy would also refer to some of the colonial pioneers, such as Émile Wangermée, the first governor of Katanga and founding father of Lubumbashi, or Alexandre Delcommune, who had been active as an officer of the *Force Publique* all over the Congolese territory in the early days of Belgian colonisation. A series of street names in the '*ville européenne*', such as the *Avenue des Chutes*, the *Avenue de la Lufira*, the *Avenue de Kambove*, the *Avenue du Zambèze* or the *Avenue du Moero* referred to geographical locations and rivers in Congo. Others seemed to serve as tools for facilitating the navigation of the city as their names were directly linked with specific urban sites, such as the *Avenue de la Gare*, the *Avenue du Cimetière* or the *Avenue de l'Abattoir*.

---

[21]Liora Bigon and Ambe J. Njoh, 'The Toponymic Inscription Problematic in Urban sub-Saharan Africa: From Colonial to Postcolonial Times', *Journal of Asian and African Studies*, 50, 1 (2015), pp. 25–40 (p. 30).

[22]Discussions within the Comité Urbain reveal that the latter decision was taken as an explicit tribute to the "beloved" Queen Astrid, who had passed away in 1935. AA, GG. 10.927 Procès-verbal réunion Comité Urbain Élisabethville, report of the meeting of the *Comité Urbain*, 16 June 1948.

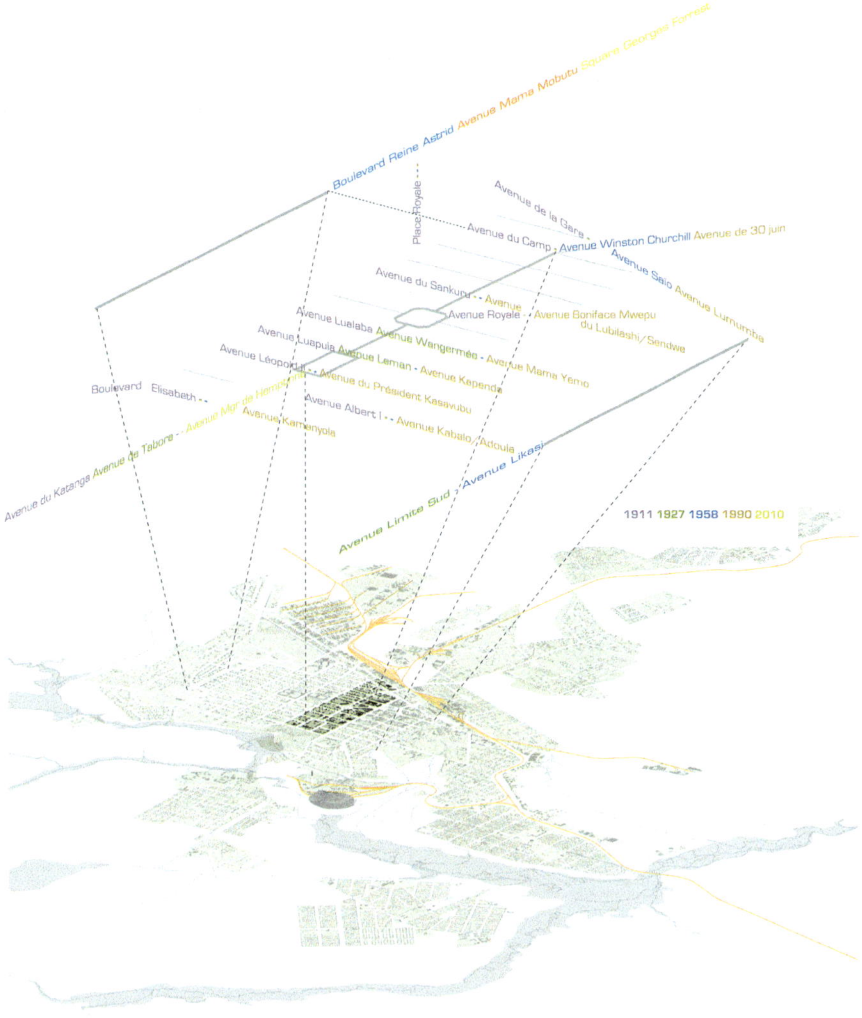

Fig. 12.3 Scheme of the evolution of street names in the city centre of Lubumbashi, map produced by Sofie Boonen, Sam Lanckriet and Daan De Vree (UGent)

Early plans of the first native town of Kamalondo indicate that, at least in theory, streets were also named.[23] All streets running perpendicular to the '*zone neutre*' received names linked to geographical locations in Congo, while those running parallel to it referred to particular ethnic groups in the African population such as the

[23]AA, GG. 15.953 'Cité indigène Élisabethville avec plans/cartes et photos', 'Plan parcellaire de la cité indigène, scale 1:2.000.'

Baluba, Basanga, Lunda, Basonge and Babemba. While this might suggest that the local colonial authorities made efforts to introduce an ethnic division in the native town, in line with the common practice in the mining camps of the *Union Minière*, research has revealed that this was not the case. What was planned, however, was a zoning according to social class. Indeed, an early plan of the *Commune Kamalondo* indicates that those areas closest to the road linking the native town to the European district were reserved for '*Noirs civilisés*' and '*Indous*', terms commonly used in colonial jargon to indicate small traders from Senegal, Sierra Leone, and from the former British India (India, Pakistan and Bangladesh). Recent fieldwork has revealed the extent to which this strategy of social stratification is still remembered today. Inhabitants of the *Commune Kamalondo* are still able of indicating precisely the houses where the so-called '*évolués*' and the West-Africans lived.[24]

If colonial authorities used street naming as a tool to inscribe power in space, one can wonder to what extent this strategy was successful and, in particular, if these street names were really incorporated in daily practices of the city's African and/or non-Belgian white population. The *Vocabulaire de la ville de Elisabethville*, a fascinating document dating from the early 1960s and presenting a personalised history of Lubumbashi as experienced by the milieu of African domestic servants, provides some clues.[25] It suggests that these servants, who, in the Katanga province, were not only working but also living in the European district, navigated the city not by referring to specific street names, but rather used buildings as visual markers and reference points along their routes or to explain someone how to find a particular location, as is suggested, for instance, by the following passage: 'Where is the house of that white person? It is where they are currently building the offices of the road traffic police, near the synagogue and the small park with flowers.'[26]

[24]This information was provided by several current inhabitants of the *Commune Kamalondo* during interviews in September 2010.

[25]For a complete commented version of the 'Vocabulaire', see Johannes Fabian (ed.), *History from Below: The 'Vocabulary of Elisabethville' by André Yav: Text, Translations and Interpretive Essay* (Amsterdam: John Benjamins Publishing Company, 1990). In line with practices in the British colonies in Southern Africa to which the Katanga province was directly linked by the Cape to Cairo railway line, African domestic personnel had permission to live in the so-called 'boyeries' at the back of Europeans' properties in the European quarter of the city. See: Ordinance of the Provincial Government of Katanga as to the '*cités indigènes*' and the 'workers' camps in the urban centres, nr. 3, 6 February 1922.

[26]'Vocabulary', IX, 5, in Fabian, *History from Below*. See also Johan Lagae and Sofie Boonen, 'Un regard africain sur le paysage urbain d'une ville coloniale belge. Architecture et urbanisme dans le "vocabulaire de ville de Elisabethville" d'André Yav', in Bogumil Jewsiewicki, Donatien Dibwe dia Mwemby and Rosario Giordano (eds), *Lubumbashi 1910–1920: Mémoire d'une ville industrielle* (Paris: L'Harmattan, 2010), pp. 107–124.

12.4 Uniting the Nation: The 'Bâtiment du 30 juin'

If colonial toponymy was, as Mudimbe argues, a way of transforming local memory, then the same holds true for the strategy of erasing this particular urban text of street names dating from the colonial era, which was a common practice in many former colonised territories. This was also the case in Congo which gained its independence on 30 June 1960. After five troublesome years of internal ethnic conflicts and attempts by foreign powers to safeguard their influence on the country in a shifting geopolitical context, Mobutu came to power in 1965. Soon, he would develop his so-called policy of a '*Recours à l'authenticité*' that was first laid down in 1967 in the so-called Manifesto of N'Sele.[27] Proposing a *recours* ('drawing upon') rather than a simple *retour* ('returning to'), Mobutu intended to define a framework for constructing a new national identity that would draw on the co-existence of the mutual advantages of modernity and tradition, while also erasing all remaining traces of the colonial legacy in order to underline "the rejection of the people of Zaïre of embracing blindly imported ideologies."[28] Culture constituted a key instrument to 'de-colonise' the nation, but so was the introduction of a new lexicon that was to have a profound impact on everyday life. The first and family names of every citizen had to be 'Africanised.'[29] To set the example, Mobutu changed his name from Joseph Désiré Mobutu to Mobutu Sese Seko Kuku Ngbendu Wa Za Banga. In a similar move, the colonial toponymy was erased from the national, regional and urban landscape. Congo became Zaïre; Léopoldville, the capital city, became Kinshasa; and almost all street names referring to the colonial past were replaced.

While Mobutu implemented this policy all over Congo, its application in Lubumbashi, the former Elisabethville, had a more specific agenda. Throughout the colonial era, the Katanga province, of which Lubumbashi is the capital city, considered itself distinct from the rest of the Belgian Congo. Claiming the province formed the economic backbone of the country, local authorities in Katanga always entertained a somewhat discordant relationship with the central administration in Kinshasa, a city they mockingly labelled '*la capitale du papier*' ('the capital city of paper'). On 11 July 1960, only a few days after Congo gained its independence, Moïse Tshombe, a local politician, declared the independence of the rich Katanga province, a move covertly supported by certain foreign powers seeking to secure their economic interests in the region. Tshombe's coup was to have a strong impact on the territory and profoundly hampered national integrity, and it was not until

---

[27] For a comprehensive discussion of Congo's politics in the post-colonial era, see Isidore Ndaywel è Nziem, *Histoire générale du Congo: De l'héritage ancien à la République Démocratique du Congo* (Paris: Duculot, 1998).

[28] Speech by Mobutu during the 28th General Assembly of the United Nations held in New York on 4 October 1973 (our translation).

[29] This policy was not only triggered by Mobutu's policy of a '*recours à l'authenticité*', but also and in particular by his profound conflict with the leader of the Catholic Church in Congo, cardinal Malula.

Fig. 12.4 Theatre building, Lubumbashi, architect Claude Strebelle & the architectural office *Yenga*, 1953–1956 (Photo by Johan Lagae, 2005)

early 1963 that the Katanga province was re-integrated into the national territory, among other reasons, due to strong pressure and military support from the United Nations.

Mobutu's policy of a '*Recours à l'authenticité*' did not only target erasing the colonial legacy. First and foremost, it was a strategy to restore national unity and define a common national identity for the country. In Lubumbashi, the renaming of one building in particular clearly testifies of this agenda: the former theatre building (Fig. 12.4). The theatre was constructed in the mid-1950s as a gift of the two major colonial enterprises in the city, the UMHK and the railway enterprise BCK, both of which celebrated their fiftieth anniversary in 1956, the year the building opened its doors. As such, it was commonly referred to as the '*Cinquantenaire*.' In 1972, however, the building was re-baptised the '*Bâtiment du 30 juin*', a name which explicitly refers to the date of DRC's independence. This renaming was far from coincidental. Indeed, the theatre was the one building in Lubumbashi that was most directly linked with Katanga's claims for independence. After having declared Katanga's independence, Tshombe installed the parliament of the new nation in this building. Its façade appeared on the new currency of the *Banque Nationale du Katanga* which he introduced. As a result, the theatre building was immediately 'politicised.'[30] Forming a physical marker of that particular episode in Congo's

[30]Donatien Dibwe dia Mwembu and Gabriel Kalaba Mutabusha, 'Lubumbashi : des lieux et des personnes', in Danielle De Lame and Donatien Dibwe dia Mwembu (eds), *Tout passe. Instantanés populaires et traces du passé à Lubumbashi* (Tervuren/Paris: Africa Museum/L'Harmattan, 2005), pp. 41–74.

history, it functions, to borrow Pierre Nora's notion, as a '*lieu de mémoire*' of Katanga's rather than Congo's independence. This connotation is actually evoked by a painting in the 1970s series *Histoire du Zaire*, a remarkable piece of 'memory work' by the Congolese painter Tshibumba Kanda Matulu, which depicts the theatre building with a caption reading 'Le 11 juillet 1961 [sic], le Katanga devient indé[pendant].'[31]

Mobutu's act of re-baptising it the '*Bâtiment du 30 juin*' should thus be understood as an attempt to counter that particular Katangese connotation and re-inscribe the building in a larger narrative of Congo as a unified nation.[32] What's more, Mobutu also renamed the streets adjacent to the building. One of the streets linking the site of this built complex to the city centre was renamed *Avenue du 30 juin*, while the main urban axis of which the theatre is the focal point, initially called *Boulevard Reine Astrid*, was re-baptised *Avenue Mama Mobutu*.[33] Mobutu's effort to wipe out the memory of a Katangese secession was furthermore reflected in the de-baptising of several streets elsewhere in the city, such as the *Avenue des Femmes Katangaises* which became *Avenue des Femmes Congolaises* or the *place Moïse Tshombe* which received its new name *Place de l'Unité nationale*. On a regional level, it was reflected by changing the name of the province from *Katanga* to *Shaba*, the former referring to a geographic region which by then had become politicised, while the latter is the Swahili word for 'copper', and as such was considered a more neutral term.

It is an irony of history that today the theatre building in Lubumbashi, after having served for a short while in 2000 as the seat of the national parliament,[34] now houses the provincial parliament of Katanga, several members of which actually continue to claim the need for a more autonomous role for Katanga. Even when

[31]In his conversation with Johannes Fabian, who documented and presented an in-depth analysis of this series in his seminal book *Remembering the Present*, Tshibumba explained the importance of the building as the "house in which Tshombe assumed power . . . and therefore he put it on a bill of Katanga money." The depiction should read 'le 11 juillet 1960', but then the *Histoire du Zaïre* should not be seen as an accurate survey of Congo's history. Its importance lies, as Fabian argues, in its capacity to present a powerful narrative of how the past is remembered via the present. Johannes Fabian, *Remembering the Present: Painting and Popular Culture in Zaire* (Berkeley: University of California Press, 1996), p. 105.

[32]In 1992, the theatre became linked to the name of Mutaka wa Dilomba, the first president of the *Assemblée de l'État Indépendant du Katanga*, yet today most people still refer to it as the '*Bâtiment du 30 juin.*'

[33]In 2010 the name of the latter avenue was changed again, to *Square Georges Forrest*, referring to a 'Belgian' entrepreneur with a long family history in Lubumbashi. Having been one of the very few foreign traders who has never left the city, even in times of great political turmoil, Forrest is a very prominent figure of today's *Lushois* society, even if he is considered a very controversial figure in Belgium – having been accused of a lack of transparency in the way he obtained some mining concessions for instance. Forrest's name was given to the avenue as a result of his investment in the refurbishment of this wide avenue, including the installing of a playground and some artwork on the green space in the middle.

[34]It was President Laurent Désiré Kabila who installed the national parliament in the theatre building in August 2000, but the initiative was short-lived.

the building in local parlance is still referred to as the '*Bâtiment du 30 juin*', its connotation with the Katangese struggle for independence seems to be lurking underneath the surface.

12.5 Between Colonial Memory and Post-Colonial Amnesia: The 'Avenue M(g)r de Hemptin(n)e'

The end of Mobutu's reign, in the literature commonly defined as '*la période de transition (1990–1997)*', already witnessed a shift in perception of the colonial past. In 1991–1992, during the so-called *Conférence Nationale Souveraine*, a dialogue was initiated between different sections of Congolese society in order to define some guidelines for the future development of the country. It was in this context that intellectuals as well as university professors, especially those trained in history, started to make a plea for reconsidering the colonial legacy. Voices were raised to re-install former colonial monuments in the public realm of Congolese cities, from which they had disappeared since the late 1960s in the context of Mobutu's '*Recours à l'authenticité*'-policy. While this re-instalment did not take place, in Kinshasa some street names referring to the colonial era were re-introduced at the initiative of a cultural commission composed of some of the nation's leading historians.[35] In 1995, the director of the National Archives, Antoine Lumenganeso Kiobe, even published a small book highlighting the main monuments and sites of DRC's capital city, including a substantial number of edifices constructed under colonial rule.[36]

This policy also had some effect in Lubumbashi. It led, for instance, to a refurbishment of the roundabout in front of the railway station, where a locomotive dating from colonial times was installed. But in terms of re-baptising streets and squares, its impact was negligible. This does not mean, however, that the colonial past has been ignored in Lubumbashi. In fact, it was in Lubumbashi rather than in Kinshasa that a very dynamic practice of 'memory work' developed in the academic milieu. Triggered by research conducted on popular culture by Johannes Fabian and Bogumil Jewsiewicki since the late 1970s, a whole series of book and exhibition projects initiated from 2000 onwards contributed to document *Lushois* urban memory, highlighting among other issues how the colonial past is remembered in local forms of painting, theatre, music and even fashion.[37]

[35]See Christine Dupont and Johan Lagae, 'Du bon usage de la mémoire coloniale: Entretien avec le Prof. Dr. Jacob Sabakinu', *Cahiers de la Fonderie. Revue d'histoire sociale et industrielle de la Région bruxelloise*, 38 (2008), pp. 42–46.

[36]Antoine Lumenganeso Kiobe, *Kinshasa. Genèse et sites historiques* (Kinshasa: Arnaza-Bief, 1995).

[37]For a synthesis of the activities of the group 'Mémoire de Lubumbashi', see Bogumil Jewsiewicki, 'Travail de mémoire et représentations pour vivre ensemble: expériences de Lubumbashi', in Danielle De Lame and Donatien Dibwe dia Mwembu (eds), *Tout passe: Instantanés populaires et traces du passé à Lubumbashi* (Paris/Tervuren: L'Harmattan/Musée Royal de l'Afrique Centrale, 2005), pp. 27–40.

Fig. 12.5 New signpost in Lubumbashi indicating 'Mr./Mgr. de Hemptinne' (Photo by Johan Lagae, 2010)

Given this important effort to recollect colonial memories of Lubumbashi's current inhabitants, how are we to understand the most recent changing of a street name mentioned in our introduction, namely the re-baptising of the *Avenue de Tabora* into the *Avenue Mgr. de Hemptinne*, an initiative that was taken by the current governor of Katanga and one of Congo's rising political stars, Moïse Katumbi Chapwe, on the occasion of the centennial of the Archdiocese of Lubumbashi in 2010? (Fig. 12.5) This particular case is all the more significant, as it concerns the street that during the colonial era constituted the symbolic axis of colonial power. Indeed, it was along this street that all the major edifices linked to the colonial authority were erected: the governor's residence, the pavilion Wangermée which constituted the first seat of colonial administration, the court of justice and the *Cercle Albert-Elisabeth*, a meeting place modelled on the English colonial club where the elite of the city met for drinks and dinner parties (Fig. 12.6). On the first urban maps, the street appears as the *Avenue du Katanga*, while a 1929 urban map already indicated it as the *Avenue de Tabora*. This shift in name was meaningful, as Tabora referred to the successful 1916–1917 campaign of the Belgian colonial army over the Germans in East Africa under the command of the general Charles Tombeur.[38] Presented in

[38]At the time, Tabora, a town in current Tanzania situated along the railway line linking Dar es Salaam to Kigoma which had been built in 1912, was the capital city of German East Africa. On 19 September 1916, the Belgian colonial forces entered Tabora, thus claiming, together with the

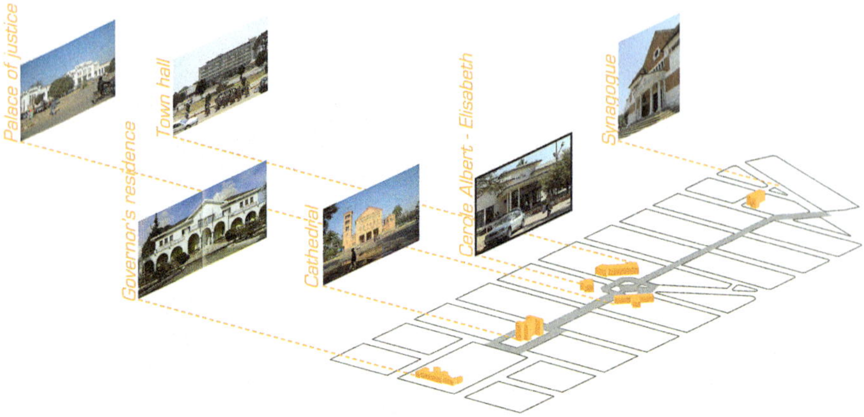

Fig. 12.6 The *Avenue du Tabora*, Lubumbashi, as an axis of Belgian colonial power, showing the main official buildings. Montage prepared by Sam Lanckriet and Daan De Vree (UGent)

national propaganda as a glorious victory, even when in reality the Germans would only be completely defeated after the end of the First World War, the reference to Tabora added a layer of national pride to this central axis of the urban grid defining the European district.[39] That Charles Tombeur served as vice-governor in Katanga between 1918 and 1920, and that his successors also had made their career in the Belgian colonial army, helps explain why the memory of Tabora, as well as that of the First World War, became so prominently inscribed in the city's urban realm. Apart from rebaptising the *Avenue de Tabora*, the local colonial authorities also had a monument to the fallen soldiers erected in 1930.

Considering this particular Belgian nationalist connotation, one can wonder why the street kept this name until 2010. In his study of Lubumbashi's toponymy, Michel Lwamba Bilonda points out that in fact during 1963 and 1971, it had been re-baptised *Avenue Munongo*, in celebration of one of the prominent figures of Katanga's independence period, but under Mobutu it had regained its initial name of *Avenue de Tabora*.[40] Yet, we can argue that at this stage it no longer

British, victory over the Germans. For the official contemporary survey of the event from a Belgian perspective, see *Les Campagnes Coloniales Belges (1914–1918)*, vol. II. 'La campagne de Tabora (1916)', Ministère de la Défense, Brussels, 1929.

[39]On 2 April 1917, the colonial army was welcomed in Lubumbashi on the return from its victorious campaign, but the renaming of the street probably took place between 1918 and 1920, when General Tombeur was residing in Katanga as vice governor general. For a critical deconstruction of the nationalist myth of Tabora, see Georges Delpierre, 'Tabora 1916: de la symbolique d'une victoire', *Belgisch Tijdschrift voor Nieuwste Geschiedenis* XXXII, 3–4 (2002), pp. 351–381; Johan Lagae, 'Lubumbashi, 2 avril 1917', in Bruno De Wever, Martine Van Asch and Rudi Van Doorslaer (eds), *Belges en guerre: Images inconnues, histoires insolites* (Waterloo: La Renaissance du Livre, 2012), pp. 30–37.

[40]Bilonda, *Histoire de l'onomastique*, pp. 47, 51.

referred to a Belgian nationalist narrative. Rather, it connected with the memory of the *Force Publique*, the colonial army. Indeed, the battle of Tabora constituted a major '*lieu de mémoire*' and a source of pride for the important *Association des Ancients Combattants*, uniting those Congolese soldiers who had participated in both World Wars.[41] Although the name remained unchanged, it now signified something completely different. Just like other streets whose name paid tribute to the *Force Publique* (such as, among others, the *Avenue Asosa*, the *Avenue Gambele* or the *Avenue Mahenge*[42]), the *Avenue de Tabora* could be used as a tool to strengthen the national unity of independent Congo.

But why then did Moïse Katumbi Chapwe decide to re-baptise the street *Avenue Mgr. de Hemptinne*, celebrating one of the most notorious figures of Katanga's colonial past? As head of the missionary congregation of the Benedictine Fathers, de Hemptinne had played a crucial role in the making and shaping of Lubumbashi. Not coincidentally, he was often referred to in contemporary sources as the 'Lion of Katanga.'[43] Was there a deliberate agenda underlying this decision, as the renaming shifted the perspective from an event of national importance (the battle of Tabora) to that of an admittedly powerful, yet local figure?[44] What was Katumbi thinking when he pronounced his decision standing on the steps of the catholic cathedral, one of the city's architectural landmarks situated on the *Avenue de Tabora*? Was he actually aware that there had already been an *Avenue de Hemptinne* at another location in the city, namely the former *Avenue du Sankuru* that connects the European district with the first native town and which today is known as the *Avenue Sendwe*? Did he ever anticipate that his decision, which he probably intended to please the church authorities, might have offended the milieu of the *Anciens Combattants*, for whom the name Tabora had so strongly resonated long after independence?[45]

[41]For a first history providing an African perspective of the *Force Publique*, see *Lisolo na Bisu 1885–1960: 'Our History': The Congolese Soldier of the 'Force Publique'*, exhibition catalogue (Brussels: Musée Royal de l'Armée, 2010).

[42]These streets are situated in the northern part of the city's industrial quarter.

[43]For a biography, see 'Hemptinne (de), Jean-Felix (1876–1958)', in *Biographie Belge d'Outre-Mer* (Brussels: Académie Royale des Sciences d'Outre-Mer, 1973), VII-A, col. 291–299.

[44]This interpretation results from a discussion between Donatien Dibwe, Johan Lagae and Serge Songa Songa, an assistant in the History Department of the University of Lubumbashi in October 2013.

[45]While there still exists an *Association des Anciens Combattants* in Lubumbashi, only a few members who served as soldiers during the Second World War still remain. A local inquiry reveals that they are today too preoccupied with their everyday struggle for survival to be concerned with the erasure of the memory of Tabora. Local memories of World War I and those of World War II often tend to get mixed up and blurred in Lubumbashi, all the more so since the former occurred at a moment in time when the African population of Lubumbashi mainly consisted of people "in transit", such as labourers engaged on short term contracts, rather than settlers. See Donatien Dibwe Mia Mwembu, 'Les souvenirs de la Grande Guerre à Lubumbashi', paper presented at the University of Kinshasa in the context of a history conference on the theme of the First World War in Congolese history, 17 November 2014.

One significant detail suggests that Katumbi was in fact improvising when making the announcement and that his decision was rather impulsive instead of being based on a profound understanding of local history and how it is passed on in *Lushois* urban memory.[46] When the first street signs signalling the new name went up, they mentioned 'Avenue Mr. Jean-Felix de Hemptine' rather than using the abbreviation 'Mgr' for 'Monseigneur', which is the official title of de Hemptinne. Soon after, the mistake was corrected, not by installing new signs, but, in a rather clumsy way, by pasting stickers with 'Mgr' to correct the wrong appellation 'Mr' (Fig. 12.6). But the misspelling of the name has remained. Even today, the sign still reads 'de Hemptine' instead of 'de Hemptinne'. In the end, Katumbi's act, then, seems far removed from Mobutu's conscious policy of erasing the colonial lexicon of Lubumbashi's toponymy or from the colonial inscription of power by renaming an African territory considered to be a *terra nullius*. Constituting an initiative fostered more by ignorance than by a deliberate and skilful strategy of transforming, or even silencing local memory, the rebaptising has not had any real impact. *Lushois* continue to refer to the street as the *Avenue de Tabora*. If there is anything that will foster the memory of Jean-Felix de Hemptinne in local memory, it is not these new street signs, but rather the impressive brick cathedral, a landmark building constructed in 1921 that still forms a crucial reference point for anyone navigating Lubumbashi's city centre.

[46]It is telling in this respect that historians of the 'Mémoires de Lubumbashi'-group were not consulted beforehand on the initiative of Katumbi.

Chapter 13
Formal and Informal Toponymic Inscriptions in Maputo: Towards Socio-Linguistics and Anthropology of Street Naming

César Cumbe

Abstract Supported by urban socio-linguistics and by an anthropological approach to toponymic inscriptions, we here analyse the discursive functioning of formal and informal place naming in Maputo. Conceived as both linguistic and social practice, toponymic (in-)formality is examined in its actual context of production and use. The official place-naming system, both colonial and revolutionary-post-colonial, emerges as a lesson in history and geography. Here the street names relate to historical events, commemorate national and foreign heroes, and pay homage to other cities and countries of the world that are closely linked to the history of Mozambique. In the unofficial case, street names serve as a sort of atlas in which, aside from names of other cities and regional localities, names of districts and rivers of Mozambique are also recalled. It should be noted that toponymic informality does not seek to replace, and much less compete, with the official naming system. On the contrary, its simultaneous existence seeks to consolidate and highlight contemporary history, together with the everyday reality and spatiality of the city dwellers. The latter includes an autochthonous linguistic heritage, collective memory, and the cultures and traditions of Mozambique. This short chapter also exemplifies that informal place naming reflects Maputo's urban multilingualism (Portuguese, English and indigenous languages). These toponymic characteristics neutralise the geographical and administrative boundaries of the urban sphere and deconstruct the centre-periphery dichotomy. This is, inter alia, through the original social functioning of transport and popular shops that exhibit place names all over the urban area, regardless of their actual socio-spatial location.

Keywords Maputo • Mozambique • Toponymic inscriptions • Lusophone Africa • Urban informality • Centre-periphery • Street signs • Popular transport

---

C. Cumbe (✉)
Faculty of Language, Communication and Arts, Universidade Pedagógica, Maputo, Mozambique
e-mail: climbe@voila.fr

L. Bigon (eds.), *Place Names in Africa*, DOI 10.1007/978-3-319-32485-2_13

13.1 Introductory Note

Supported by urban socio-linguistics and by an anthropological approach to toponymic inscriptions, we here analyse the discursive functioning of formal and informal place naming in Maputo.[1] Conceived as both linguistic and social practice, toponymic (in-)formality is examined in its actual context of production and use. The official place-naming system, both colonial and revolutionary-post-colonial, emerges as a lesson in history and geography. Here the street names relate to historical events, commemorate national and foreign heroes, and pay homage to other cities and countries of the world that are closely linked to the history of Mozambique. In the unofficial case, street names serve as a sort of atlas in which, aside from names of other cities and regional localities, names of districts and rivers of Mozambique are also recalled. It should be noted that toponymic informality does not seek to replace, and much less compete, with the official naming system. On the contrary, its simultaneous existence seeks to consolidate and highlight contemporary history, together with the everyday reality and spatiality of the city dwellers. The latter includes an autochthonous linguistic heritage, collective memory, and the cultures and traditions of Mozambique. This short chapter also exemplifies that informal place naming reflects Maputo's urban multilingualism (Portuguese, English and indigenous languages). These toponymic characteristics neutralise the geographical and administrative boundaries of the urban sphere and deconstruct the centre-periphery dichotomy. This is, inter alia, through the original social functioning of transport and popular shops that exhibit place names all over the urban area, regardless of their actual socio-spatial location.

13.2 Maputo's Geographical Division of the Urban Area: Socio-Spatial Inequality

Mozambican geographers normally divide Maputo City into centre, suburbs, urban outskirts, and agriculturally worked savannah and fields.[2] This spatial delimitation not only "highlights the historical development and the differentiation of the city into functional areas",[3] but also reflects the geographical and sociological setting of the Maputo urban area. The centre hosts prestigious neighbourhoods of colonial

---

[1]This chapter was translated from Portuguese into French and then into English by Carlos Mataruca, Universidade Pedagógica, Faculdade de Ciências da Linguagem, Comunicação e Artes, Maputo, Moçambique, and the Editor.

[2]Aniceto dos Muchangos, 'Traços gerais da cidade de Maputo', *Boletim do Arquivo Histórico de Moçambique* n°2, Outubro 1987, pp. 122–135; Maria Clara Mendes, *Maputo Antes da Independência: Geografia de uma Cidade Colonial* (Lisboa: Centro de Estudos Geográficos da Universidade de Lisboa, 1979).

[3]Dos Muchangos, 'Traços gerais da cidade de Maputo', p. 122.

origin where mainly the elites live. The suburbs, characterised by informal habitats and social precariousness, host old neighbourhoods that were intended, inter alia, for Indian workers and traders in colonial times. During the Portuguese colonial period (till 1975), the socio-spatial occupation was based on economic and racial segregation criteria. In the suburbs, close to the worked savannah, food is produced to feed the markets in the city centre. The urban periphery, this characterised by insufficiency in urban infrastructure,[4] hosts neighbourhoods distant from the city centre and the ones most affected by unemployment. It borders on the countryside.

However, there is no consensus among geographers about the delimitation of Maputo's urban area.[5] The suburbs and the periphery are generally considered a homogeneous area versus the city centre. This fluctuation of boundaries is due to the fact that the majority of the population living in the first two urban areas mentioned essentially lives off informal-sector activities. In fact, travelling from the south to the north of Maputo, the growing socio-spatial inequalities are notable as we move away from the centre, through the suburbs, and then the urban periphery, to reach the agrarian land. To mention a particular example, the streets of the city centre are all paved, listed on city maps, and almost all have been given official names. How therefore shall we render the toponynic inscription of the whole urban complex, especially the street names? Are there socio-linguistic variations between the centre, the suburbs and the urban periphery? Before answering these questions, the administrative and toponymic division of Maputo City needs to be presented.

13.3 Maputo's Administrative Divisions and Districts: The Return of Indigenous Names

Administratively, metropolitan Maputo consists of two capitals: the capital city, namely Maputo City; and the capital of Maputo Province, Matola City. The geographical and administrative divisions of the urban area of Maputo differ slightly. While the geographical division normally divides the city into four zones (centre, suburbs, urban periphery and rural periphery), the administrative one divides the city

[4]This is in terms of drinking water, sanitation, electricity and public lighting, garbage collection, public spaces and social services such as health centres, banks, markets, shops, police stations, leisure centres and adequate places of worship. See, for more: José Forjaz (et al.), *Moçambique, Melhoramento dos Assentamentos Informais, Análise da Situação & Propostas de Estratégias de Intervenção* (Maputo: Centro de desenvolvimento do Habitat (CEDH), 2006), p. 42.

[5]The lack of consensus among geographers is due to the heterogeneity of the variables considered in the categorisation of these neighborhoods. While some geographers tend to consider the socio-demographic variables, urban infrastructure and housing types, others use a multidimensional analysis of the diversity of habitation and land use categories, as were obtained in the 1997 census through information systems and remote sensing . In this case, aside from the city centre, Mozambican geographers identify four categories of residential space in Maputo, i.e., former suburban areas, and recent peri-urban, urban and rural margins. These fluctuations clearly show the difficulties of socio-spatial delineation of the city.

Table 13.1 Municipal Districts and attached toponyms (2001, 2010)

Municipal districts	Neighbourhoods
KaMpfumo	Alto-Maé 'A' and 'B'; Central 'A', 'B' and 'C'; Coop; Malhangalene 'A', 'B' and 'C'; Polana Cimento 'A' and 'B'; Sommerschield
Nhlamankulo	Aeroporto 'A' and 'B'; Chamanculo 'A', 'B', 'C' and 'D'; Malanga; Minkadjuine; Munhuana; Unidade 7; Xipamanine
KaMaxakeni	Maxaquene 'A', 'B', 'C' and 'D'; Mafalala; Polana Caniço 'A' and 'B'; Urbanização
KaMavota	Ferroviário; 3 de Fevereiro; F.P.L.M.; Hulene 'A' and 'B'; Laulane; Mahotas; Mavalane 'A' and 'B'
KaMabukwana	Bagamoyo; George Dimitrov; Inhagóia 'A' and 'B'; Jardim; 25 de Junho 'A' and 'B'; Luis Cabral; Magoanine; Malhazine; Nsalene
Ka Tembe	Ilha de Catembe
KaNyaka	Ilha de Inhaca

into seven urban districts, five of which are continental and two are islands. These administrative districts were entitled 'Municipal Districts' in 2001, and renamed with indigenous names in 2010 as shown in Table 13.1.

This spatio-administrative ordering of the current city expresses the will of the authorities to fight the deep vestiges of racial and social segregation, inherited from the colonial city. However, in the everyday language of Maputo periphery residents, the city centre is designated as 'Xilonguine' (that is, 'the place of the white people'), and every inhabitant of the city centre is called 'Mulungo', which in the local Changana language means 'white person.' The rural people, use the term 'Xilonguine' thus without making a distinction between the centre and the periphery. Consequently, the term 'Xilonguine' now covers two spatialities: the historic colonial town as opposed to the peri-urban city; and the current city including its centre and periphery as opposed to the rural area.[6]

13.4 Maputo's Formal Street Names: An Open-Air Lesson in History and Geography

Assuming that "the street names constitute a history lesson that is learnt relatively easily by those who do not have much time"[7]; and that their inscription is "the most obvious [though selective and somewhat authoritarian] reading of the city"[8] – the urban text of the streets of Maputo arouses considerable interest. The names of

[6]Isabel Raposo, 'Décentralisation et plans d'aménagement de quartiers péri-urbains à Maputo', in Laurent Fourchard (ed.), *Gouverner les villes d'Afrique: Etat, gouvernement local et acteurs privés* (Paris: Karthala, 2007), pp. 53–68 (p. 55).

[7]Raposo, 'Décentralisation et plans d'aménagement', p. 55.

[8]Raposo, 'Décentralisation et plans d'aménagement', p. 55 (brackets added by author).

the main streets of the city centre, both colonial and revolutionary, make reference to historical events of the city and of the country, such as: 'Avenue 10 November', 'Avenue 24 July' or 'Avenue 25 September' (respectively: day of the city of Maputo, independence day, day of the struggle for national liberation). They also seek to honour international socialist or communist martyrs and heroes of the country, such as: 'Avenue Friedrich Engels' and 'Avenue Karl Marx', 'Avenue Alberto Luthuli' (after the ANC president and 1960 Nobel Peace Prize Laureate, 1898–1967), 'Avenue Ho Chi Minh' (after the leader of Communist Vietnam, 1920–1969), 'Avenue Eduardo Mondlane' (the first president of FRELIMO, the Mozambique Liberation Front, murdered in 1969) and 'Avenue Filipe Samuel Magaia' (the founder of FRELIMO, killed in 1969).[9] Other names refer to cities and foreign countries related to the history of Mozambique. The dominant toponyms that recall foreign cities in the city centre are fundamentally of Portuguese cities, namely: Alba, Alentejo, Algarve, Aveiro, Beja, Braga, Bragança, Castelo Branco, Dão, Évora, Gávea, Guarda, Seia and Minho. Interestingly, all these streets are located in the same neighbourhood, Malhangalene 'B'. The names of Portuguese cities are followed by names of several Tanzanian cities, namely: Bagamoyo, Dar-es-Salaam and Nachingwea. While the names of Portuguese cities echo the colonial history of Mozambique, the names of the Tanzanian cities echo the history of the anti-colonial liberation struggle in Mozambique, in which Tanzania played a crucial role. As for names that evoke foreign countries, we shall highlight Angola, Algeria, Guinea-Bissau, France, State of Goa, East Timor, Tanzania, Zambia and Zimbabwe.

While certain names provide a history lesson of Mozambique, others assert a lesson in Mozambique's geography. In the city of Maputo these predominantly consist of names of cities, districts and rivers of the other provinces of Mozambique. The prominence of such a 'geographical atlas' in miniature is also observable in the peri-urban neighbourhood of Munhuana, where twelve place names and five names of rivers from this country are to be found.[10] In addition, the ethno-history of Southern Africa[11] reveals that the abundance of indigenous place names has to do with the colonial policy of grouping indigenous people according to race and ethnicity in the colonial town of Lourenço Marques (now Maputo) in the 1890s.

---

[9]Raposo, 'Décentralisation et plans d'aménagement', p. 55. Catherine Farvacque-Vitkovic, Lucien Godin, Hugues Leroux, Florence Verdet e Roberto Chávez, *O endereçamento e a gestão das cidades* (Washington: Banco Internacional para Reconstrução e Desenvolvimento/ Banco Mundial, 2005), p. 7.

[10]César Cumbe, 'Maputo, là où parlent les murs et les voitures', Actes du Colloque (conference proceedings) entitled 'Cultures urbaines à la Réunion', Saint-Denis, 13–14 avril 2011.

[11]Farvacque-Vitkovic (et al.), *O endereçamento e a gestão das cidades*, p. 7; Valdemir Zamparoni, 'Monhés, Banianes, Chinas e Afro-maometanos: Colonialismo e racismo em Lourenço Marques, Moçambique, 1890-1940', *Lusotopie* (2000), pp. 191–222; João Craveirinha, 'Thomas Mapfumo' in http://macua.blogs.com/moçambique_para todos/thomas_mpfumo.doc (visited 30 June 2015).

These urban toponyms recalled the indigenous people's places of origin, as in the ideology of the colonial state, the indigenous sector was generally considered not more than a cheap labour force without any legal standing. After one hundred and twenty five years, this urban text, which still survives, recalls this cruel reality, stored in the memory of these neighbourhoods. The colonial spatial stratification that was marked by a "territorial separation of bodies, hierarchy of race and colour"[12] was prominent in the occupation of urban areas: the downtown area (offices, business, leisure and pleasure), the upper town (an exclusive area for whites) and the suburbs (area for the blacks). Such a hierarchy of race and colour was also felt in the use of public transport and shops.

While going through the official place names of Maputo, we can find, however, names that refer to institutions such as: the OAU (Organisation of African Unity), the UN, archbishops, church, mosque, press, radio and factories. In addition to the colonial and revolutionary history, the geographical atlas of settlements and rivers is to be found. Aside from further names advertising institutions and infrastructure, we find a variety of street names that include generic adjectives and specific nouns, which could be classified into: university degrees (Dr, diploma); profession (engineer, poet, actor and actress); religion (bishop, father, our lady and compassion); military (general, commander, lieutenant, base, military school and major); political hierarchy (president, dom, dona); social etiquette (Street of Blood Donor, Street of Volunteers, Street of Angola's Heroes); and the field of botany and agriculture (plants, fruits, flowers and seasons).

13.5 Maputo's Informal Street Names: An Account of Contemporary History and Daily Life

Spontaneous names in the suburbs allow city dwellers to express themselves by the social act of inscribing toponyms. In these urban areas, where informal bottom-up naming is very common, both fixed or permanent signposts and temporary or mobile signposts[13] make sense for the space users. In contrary to the city centre, where each street has one descriptive name, in the suburbs official names coexist with informal names. The official system frequently consists of a numbering system, while the informal names are more meaningful for the residents; one officially numbered street can sometimes have several unofficial names. More than just names, the unofficial names have many functions or meanings, which are not necessarily linked to place names proper. For instance, these names may warn against the danger of theft or rape – such as 'Cross Running Street', 'Armed People Street'; they can

[12]Farvacque-Vitkovic (et al.), *O endereçamento e a gestão das cidades*, p. 7.

[13]See next sub-heading for an example for mobile signs.

report the gossip of the neighbourhood – such as 'Street of Gossipy Mothers', or ironically 'BBC Street'; they can suggest intimate encounters, e.g., 'Valentine Street', 'Passionate Street'; they can act as informal transport network indicator, e.g., 'Stop Circus Street', 'First Street Stop'; and they may enhance the inhabitants of the neighbourhood through the appropriation of names of other streets from prestigious neighbourhoods ('Gil Vicente Street'). Such names can also reveal a sense of humour amongst the everyday space users. For example, in the peripheral district kaMavota, the street '4.821' (official name), is known informally also as 'Avenida Gato Bravo', that is, 'Wild Cat Avenue.' This is because initially, the following announcement was advertised on an unofficial public notice-board: 'A Venda Gato Bravo' – that is, 'a wild cat for sale.' After a while, someone added an 'i' after the 'n' in the word 'A Venda', and by thus it turned into 'Avenida', which means 'Avenue.' As we can see, the informal street names of the suburbs can act as a gateway for capturing the actual life of the urban dwellers, attached facets and trajectories (Figs. 13.1 and 13.2).

Fig. 13.1 'Rua 14.205' (Street 14.205) and 'Paragem Rua do Círculo' (Square Street Station).Photo illustrates the harmonious co-existence of formal and informal place names in the suburban neighbourhood of Zikwama in Matola (the capital of Mapoto Province) (Author's photo)

Fig. 13.2 The junction of 'Rua 5.765' (Street 5.765) with 'Rua 5.750' in KaMavota, a suburban neighbourhood of Maputo, Mozambique's capital. These streets are also informally called 'Rua dos Apaixonados Cont'/'Hélio Eles' (Streets Lovers Contact/ They Hippolytus) (Author's photo)

13.6 Between Itinerant and Fixed Place Names: Geographic and Socio-Linguistic (Dis-) Continuities

Maputo's municipal authorities require each privately-owned vehicle of unofficial public transport, or minibus, to specify the route with coloured banners.[14] Four banners are required (one at the front, one at the rear and one on each side), indicating the departure and destination points.[15] These must correspond with the

[14]Private public transport in Maputo and Mozambique in general are called 'Chapa 100.' Initially, this designation referred to pick-up trucks fitted with benches and covered with zinc to protect passengers against the sun and rain, hence the name 'chapa' to indicate the zinc roof and '100' to indicate the fare (100.00 meticals, the local currency). Gradually these vans have given way to the minibus (Toyota Hiace especially). But considering the current transport crisis, the pick-ups have come to rely on minibuses, this time without either bench or zinc roof. Passengers are shaken and have to stand, often close to each other like lovers, hence the new name 'My Love', very common for these minibuses, mainly in Maputo.

[15]Each minibus has four banners of the same colour showing the way. The colours are found, inter alia, red, yellow, blue and green. A certain colour is used to indicate the routes that are in the same direction or geographical area. Yellow, for instance, covers the following routes: Museu (city centre)-Malhazeine (periphery); Museu (city centre)-Albazine (periphery); Museu (city centre)-Hulene (periphery); Museu (city centre)-Laulane (periphery); Baixa (downtown) -Praça

names of the neighbourhoods served. Thus the names of the more prestigious districts and the names of popular neighbourhoods that are served by the unofficial transport appear everywhere side by side. This linguistic landscape transforms the vehicle into a new territorial paradigm, a 'living laboratory'[16] of conjunct spatial associations in a variety of languages. Most of these urban toponyms are African (such as: Cingatela, Congolote, Lhavela, Mussumbuluco, Malhapswene, Mahotas, Mikadjuine, Matendene, Zikwama and Tsalala), but European toponyms are noticeable as well (such as: Drive In, Sommerschield, Jardim, Central, Liberdade, Costa del Sol, Fomanto, Cidade da Matola, Benfica, and Zona Verde). They expose the way the vernacular languages are spoken by most of the city dwellers and equally the foreign languages, of which the Portuguese language occupies a central place. The names of the series of informal stops are usually descriptive and may indicate the existence of something or someone in their proximity. Revealing the quick-witted creativity of the city dwellers, these names might include a take away, neighbourhood, street, tree, bridge, square, business office, hospital, cemetery, dumping ground, school or other landscape mark.

An examination of the informal signposting reveals multilingual announcements or non-linguistic signs such as obsolete objects or slabs of official road signs ripped from the city centre, which are used mostly for bus stops outside the city centre. The names of take-away restaurants are usually re-appropriated in the everyday oral discourse in the distant suburbs as shops, popular transport stops and points of orientation. This is in order to locate family members, friends or 'invisible' offices of informal service providers, such as traditional healers, dog trainers, builders, plumbers, teachers or tutors. The advertising plates of the latter providers are exposed in every possible site: hanging on trees, poles, fences, and walls. Even visitors coming from rural areas orient themselves quite easily in the city, and especially in its periphery, thanks to the names of the popular takeaways (Figs. 13.3 and 13.4).

The itinerant place names in Maputo's public transport reveal a city in motion that challenges and deconstructs the rigid geographical and administrative boundaries. In fact, Maputo's history, culture and architecture cannot be appropriately understood without a reference to the toponyms of its streets and neighbourhoods, as well as without a reference to the names of its inhabitants, that is, the indigenous anthroponyms. These fixed-cum-mobile toponymic inscriptions function as a kind of the 'DNA of the city',[17] enabling one to grasp the actual city's complexity and complicity, its spontaneity, intelligence, authenticity and anonymity. They also

back Combatentes (periphery); Anjo Voador (city centre)-Praça back Combatentes (periphery). Note that the use of the same colour banners here is logical because the periphery includes neighbouring areas that follow one another in sequential order, e.g., Praça dos Combatentes, Laulane, Magoanine, Mlalhazine, Albazine.

[16]Giovanni Brino, 'La signification urbaine: dichotomie entre centre ville et périphérie en Italie', in Bernard Lamizet (ed.), *Les langages de la ville* (Paris: Éditions Parenthèses, 1997), p. 59. See also: Raposo, 'Décentralisation et plans d'aménagement', p. 55.

[17]Gilles Rabin, Luc Gwiazdzinski, *Si la ville m'était contée* (Paris: Eyrolles, 2005), p. 69.

Fig. 13.3 Photo illustrates the colour banners required by the municipal authorities to indicate the route of public transport: Museu (city centre)-Magoanine (periphery), and the informal inscription 'Grande Xipila' (Big Xipila) which means 'big boss' (in a combination of Portuguese and Changana) (Author's photo)

Fig. 13.4 Photo illustrates the colour banners of: Museu (city centre)-Malhazine (periphery) and the informal inscription 'Nhimpine' which in Changana means 'in the war.'

emphasise that Maputo is a city open to the free movement of people and goods, enabling its residents a daily 'reading' throughout the disorder. Finally, Maputo is both a city of letters and a capital of spatial signs. Its inscriptions play a crucial role in the material culture, operating in multifunctional social areas (e.g., residential, commercial, catering[18]) and as symbolic expressions (e.g., historiologies, events, memories, monuments). Maputo, by attracting a movement of itinerant players, remains a prominent place of reference for the coexistence of official and unofficial place names, a reservoir of collective memory recounting the city and the country in their different facets.

13.7 Conclusion

In this chapter we studied the toponyms as an area of linguistic practices (enumerations, words, phrases, texts, oral corpuses) and as a reference of identity, exposing urban multilingualism and marking territories of spatial symbolism and control on the part of daily peaceful powers. The names of neighbourhoods, streets, avenues, squares, markets, and popular shops and transport reflect stories that immerse us in the city of Maputo, conceived as a vivid "active social laboratory."[19] Moreover, the obligation to use colour banners by Maputo's municipal authorities on local public transport results in the appearance of prestigious place names side by side with those of peripheral neighbourhoods, and thus contributes to the everyday integrity and visibility of the entire conurbation. This transport also circulates freely in the urban space as an exponent of continuum, despite the municipal and perceptual divisions of the city.[20] Maputo's formal and informal place names make us aware that it is a city that speaks (and not one that is talked about). The city lets us browse into its discursive and textual toponymy like reading a book that is written by multiple hands, trying to codify and de-codify the multiplicity of its social and individual messages.

[18]Today in Maputo, the classic restaurants offering their daily dishes are open towards the street. They are faced with competition from restaurant-cars operating at every corner of the busiest streets in the heart of the city, at lunch-time serving different take-away dishes at low cost.

[19]This expression is borrowed from the Chicago School and was born in the 1920s following a fascination with human behaviour in urban environment. Robert Park, Ernest Burgers, Roderick McKenzie and Louis Wirth, are among the pioneers who laid the foundation for an ecological approach to the city.

[20]Christian Topalov (ed.), *Les divisions de la ville*, collection 'les mots de la ville' (Paris: UNESCO & Maison des sciences de l'homme, 2002); Bernard Lamizet and Pascal Sanson (eds), *Les langages de la ville* (Marseille: Parenthèses, 1997).

Chapter 14
Glocal Naming and Shaming: Toponymic (Inter-)National Relations on Lagos and New York Streets

Wale Adebanwi

Abstract Streets are sites of hegemony and counter-hegemony, of inclusion and exclusion, of incorporation and expulsion, and of cooperation or conflict. Thus, in the cultural geography of cities, commemorative street names are critical sites of social reproduction. Recent literature on toponymy calls attention to such practices as important cultural and political arenas for understanding socio-political processes, but often focuses on the politics and sociality of street naming within local, national politics to the exclusion of how local politics intersects with international politics. This chapter examines the politics of spatial inscription and the social reproduction of 'place' or 'space' on a street corner in New York City named after Kudirat Abiola, an assassinated woman activist in Nigeria, and the retaliatory renaming by the military regime of a Lagos street hosting the US Embassy after the African American anti-establishment activist Louis Farrakhan. Subsequently, the next democratic government of Nigeria renamed the street, this time after the US ambassador, the African American Walter Carrington. Toponymy, the chapter concludes, can thus be seen as a form of retortion in international relations.

Keywords Street names • Lagos • New York • International politics • Toponymic retortion • Abiola • Louis Farrakhan

14.1 Introduction

Streets are sites of hegemony and counter-hegemony, of inclusion and exclusion, of incorporation and expulsion, and of cooperation or conflict. Thus, street naming is a potential manifestation of different socio-political dynamics – both discursive

W. Adebanwi (✉)
African American and African Studies, University of California, Davis, CA, USA
e-mail: anadebanwi@ucdavis.edu

L. Bigon (eds.), *Place Names in Africa*, DOI 10.1007/978-3-319-32485-2_14

and material.[1] In the cultural geography of cities, commemorative street names are critical sites of social reproduction. They also carry the burden of specific forms of history, politics, memory and modes of remembering and affirming identity. However, even though recent literature on toponymy has called attention to reading such practices as important cultural and political arenas for understanding socio-political processes, it often focuses exclusively on the politics and sociality of street-naming within local and national politics. In fact, most scholars writing on the street see it strictly as 'specific, local landscapes manifest[ing] broader social and cultural processes.'[2] The 'specific and local' nature of the street is hardly conceived in terms of its insertion in international socio-cultural, economic and political processes; and, for the most part, the scholarly literature on street naming has not paid deserved attention to how local politics sometimes intersects with international politics. An exception to this is Reuben Rose-Redwood's essay on the renaming of the Sixth Avenue as the Avenue of the Americas in post-World War II New York.[3] Against the backdrop of the debate among political geographers on the political construction of scale, Rose-Redwood has explored the 'performativity of street naming' in demonstrating how street naming can be used 'as a strategy for projecting the imaginative geographies of geopolitics onto the urban streetscape, thereby remaking the "local" landscape as a metonymic representation and performative enactment of "supranational" relations.'[4] What do such practices, both in their discursive and material dimensions, mean for our understanding of glocal (combining the local and the global[5]) and (inter-)national politics?

This chapter addresses this question by attempting to understand how particular communities (local, diasporic and transnational) define the 'local' and consequently link different scales 'by symbolically associating the fate of the "local" with that of the "supranational"',[6] consequently producing critical political implications not only for the 'political construction of geographic scale'[7] but also for the global

[1]An earlier version of this chapter was presented at the Canadian African Studies Association Conference, York University, Toronto, Canada, 5–7 May 2011. A longer version was published in *African Affairs*, 111/455 (2012) – reprinted by permission on behalf of Oxford University Press.

[2]Nicholas Fyfe, 'Introduction: Reading the Street', in Nicholas Fyfe (ed.), *Images of the Street: Planning, Identity and Control of Public Space* (London, New York: Routledge, 1998), pp. 1–12 (p. 1).

[3]Reuben Rose-Redwood, '"Sixth Avenue is Now a Memory": Regimes of Spatial Inscription and the Performative Limits of the Official City-Text', *Political Geography*, 27, 8 (2008), pp. 875–894.

[4]Rose-Redwood, '"Sixth Avenue is Now a Memory"', p. 878.

[5]The term 'glocal' is used to point to the 'increasingly dense superimposition and interpenetration of global political-economic forces and local-regional responses within the parameters of a single, re-scaled framework of state territorial organization.' See Neil Brenner, 'Global Cities, Glocal States: Global City Formation and State Territorial Restructuring in Contemporary Europe', *Review of International Political Economy*, 5, 1 (1998), pp. 1–37.

[6]Rose-Redwood, '"Sixth Avenue is Now a Memory,"' p. 876.

[7]Rose-Redwood, '"Sixth Avenue is Now a Memory,"' p. 876.

politics of human rights, liberty, and democratic self-determination. It focuses on street naming – and shaming – in two global cities, Lagos and New York, elaborates and illustrates 'the intertextuality of spatial inscription as well as the relationality of place making.'[8] Such a relational analysis of street naming facilitates our understanding of how naming is imbricated in symbolic struggles and how such symbolic struggles are also enmeshed in material struggles. Moreover, the social process of knowledge production, and the politics that is imbricated in this, can be used not only for *naming* (including describing and evaluating) what is important in local-national and international contexts, but also in *shaming*.

This chapter also aims at making a contribution to both the theoretical and empirical studies of toponymy by explicating how a recasting of governmentality as 'the conduct of misconduct' can shed new lights on our analysis of the political and ideological instrumentalisation of street naming; and by highlighting the contributions that the expanding literature on toponymy can make to the study of international relations in terms of symbolic foreign policy or symbolic international politics (particularly regarding human rights violations). In the context of the latter, I argue that when street names are targeted at foreign governments/states, they potentially constitute retaliatory measures, or retortion in international law. Another aim is to examine particular examples of how hegemonic and counter-hegemonic struggles in the context of democratisation struggles in Africa are enacted over street naming across borders. I examine the politics of spatial inscription and the social reproduction of 'place' or 'space' in a corner-street in New York, named after Kudirat Abiola, an assassinated woman activist in Nigeria, and the retaliatory renaming by a military regime of a Lagos street hosting the US Embassy after the African American anti-establishment activist, Louis Farrakhan. This was followed by yet another renaming of the same street by the succeeding democratic government after the US Ambassador to Nigeria, the African American Walter Carrington. Research for this chapter draws on formal and informal interviews in Nigeria and the United States, as well as media reports in Nigeria, United States and United Kingdom.

The Foucauldian concept of governmentality is relevant here in understanding the ways in which the discourses and materiality of toponymy – and the forms of interpellation which they bring to place – are enfolded into the tensions between subjectivity and power. While Foucault, in engaging with the analytics of power, describes governmentality as the 'the conduct of conduct' – 'an assemblage of practices, techniques and rationalities for the shaping of the behaviour of others and of oneself'[9] – I draw attention to particular forms of governmentality (in

[8]Reuben Rose-Redwood, 'From Number to Name: Symbolic Capital, Places of Memory and the Politics of Street Renaming in New York City', *Social and Cultural Geography*, 9, 4 (2008), p. 431–452 (p. 433).

[9]Michel Foucault, 'Governmentality', in Graham Burchell, Colin Gordon, and Peter Miller (eds), *The Foucault Effect: Studies in Governmentality* (Chicago: University Of Chicago Press, 1991), pp. 87–104; Mitchell Dean, *Governmentality: Power and Rule in Modern Society* (London: Sage Publications, 2009), p. 250.

a far narrower sense than Foucault intended) which manifest, in both discursive and material terms, as the *conduct of* (assumed or presumed) *misconduct*. By the conduct of misconduct, I refer to those discourses, actions and practices that seek, in the context of local, national and international relations, to structure, define, determine, legitimise or de-legitimise the discourses, actions and practices of adversaries engaged in particular contests over both symbolic and material goods and relations. Through discursive or symbolic action – such as street naming and the shaming that specific naming may articulate, denote or connote – agents, as protagonists in relational struggles, can seek to determine how their antagonists are constructed or deconstructed and the 'proper' or 'improper' ways to read preceding, existing or succeeding retaliatory actions.

In my understanding of the conduct of misconduct within (inter-)national relations, street (re)naming can provide opportunities for symbolic retortion – the retaliatory gesture which, in international law, is described as a non-amicable action, short of war, which one state takes against the other in response to conduct which the retaliating state considers injurious or unfriendly. The original 'injurious' act, and the subsequent injurious act, do not have to be provoked by actions which are directly injurious to the acting agents; it is sufficient that they are considered injurious to issues, interests, people, groups, movements, or ideas regarded by the acting agents as critical in their self-constitution and their articulation of common humanity or shareable global values – such as democracy, freedom, justice, equity, and equality.

14.2 The Toponymy of Liberty: Kudirat Abiola Corner in New York City

In the morning of 4 June 1996 in Lagos, Nigeria, a fiery woman activist, Kudirat Abiola, was driving from her Lagos mainland home to the Lagos Island to keep an appointment with the Canadian High Commissioner when she was assassinated by gun men suspected, at that time, to be members of the assassination squad of General Sani Abacha, Nigeria's dictator.[10] Kudirat Abiola was the most senior wife of Moshood Abiola, the billionaire business man and publisher, who had won the presidential election conducted on 12 June 1993. Abiola's internationally recognised election victory was annulled by the military regime of General Ibrahim Babangida on spurious grounds, and he was later jailed by the succeeding Abacha regime for protesting against the annulment and declaring himself the president in June 1994.

After Moshood Abiola's arrest and imprisonment, the pro-democracy coalition which formed to campaign for the restoration of his mandate and re-democratisation in Nigeria had found an unusual ally in Kudirat Abiola, in what was an otherwise

[10] Aminu Tijani, 'The Terror Gang', *TELL* [Lagos], 27 December 1999, pp. 14–18.

large and conservative family.[11] Kudirat Abiola had become the face of the pro-democracy movement, speaking at rallies, meetings, courts and mobilising Nigerians and foreign governments and supporters of the democratic movement in Nigeria to ensure the termination of General Abacha's autocracy and the restoration of her detained husband's mandate. As a vocal woman and rallying point, the regime realised that threats and actual imprisonment and harassment of opposition elements were not sufficient to deter her and many other pro-democracy activists. The regime therefore, decided on physical elimination as deterrence to the activists.[12] At this point, the regime's repressive measure had already led to the exile of key pro-democracy activists. Some other activists who stayed back in Nigeria either went underground or continued to risk their lives in opposition to the increasingly homicidal regime.

Kudirat Abiola's assassination shocked both local and international observers. Many key foreign governments, including the United States, Britain, Canada, and South Africa, joined Nigeria's pro-democracy activists in condemning the assassination and calling for an investigation. By the time of the assassination, General Abacha – who seemed to have particularly reviled the United States – had become an international nuisance of no small measure. Yet, the United States did not know what to do with him given Nigeria's strategic position in West Africa and her vast oil resources.[13] However, many Nigerian political and economic exiles in the UK, US and Canada were pressuring these countries to do something concrete and direct about General Abacha. While this was going on, in November 1995, the Abacha regime executed the writer and minority rights activist, Ken Saro-Wiwa, and eight other Ogoni environmental-rights activists. This led to the suspension of Nigeria from the Commonwealth.

Against this backdrop, Nigerian activists in the US and their American sympathisers proposed renaming the north-east corner of New York's Second Avenue and 44th Street, where the 22-storey Nigerian Permanent Mission to the United Nations is located, in recognition of Ken Saro-Wiwa. They were still considering this in June 1996 when Kudirat Abiola was assassinated. At this point, Jumoke Ogunkeye, a New York-based Nigerian-American and Chairman of United Committee to

[11]Abiola had several wives and many children.

[12]Abraham Useh, 'The Assassins', *TELL* [Lagos], 7 September 1998, pp 12–19; Ade Olorunfewa, 'Abacha's Murder Incorporated', *TELL*, 5 October 1998, pp. 10–15; Dele Agekameh, 'The Story of a Killer', *TELL*, 20 December 1998, pp. 14–20. General Abacha's Chief Security Officer and head of his personal security outfit, 'Strike Force', Major Hamza Al-Mustapha and Lateef Shofolahan, Kudirat Abiola's personal aide, were both sentenced to death by a Lagos High Court on 29 January 2012, for the murder of Kudirat Abiola. Both have appealed the sentence. See: 'Al-Mustapha, Shofolahan Sentenced to Death by Hanging', *Nation* [Lagos], 30 January 2012, p. 1 and 'Death Sentence: Al-Mustapha Files Appeal', *Daily Trust* [Abuja], 31 January 2012, p. 1.

[13]Nigeria is the fifth largest exporter of crude oil to the United States, accounting for about fifty percent of all crude exports from the country and about ten percent of total US oil imports. Paul Lubeck, Michael Watts, and Ronnie Lipschutz, 'Convergent Interests: U.S. Energy Security and the "Securing" of Nigerian Democracy', *International Policy Report*, February (2007), pp. 1–23.

Save Nigeria (UCSN), who also worked at the New York City Council, decided to mobilise his contacts within the New York City to ensure the maximum embarrassment possible for the Abacha regime, by getting the Council to rename the corner of the Nigerian Mission to the UN after Kudirat Abiola.[14] Ogunkeyede was joined by Michael Fleshman, human rights coordinator for the Africa Fund, a US non-profit organisation. In July 1996, both helped to persuade the New York City Council to pass a resolution condemning the Abacha regime and the killing of Kudirat Abiola. In May 1997, the Community Board Six voted to ask the City Council to put Kudirat Abiola's name alongside the existing street signs in front of the Nigerian consulate at 828 Second Avenue, by 44th Street. That way, Ogunkeyede stated, Nigeria's consulate employees 'would have to look at it every day.'[15] The Abacha regime reacted strongly against the move, and deployed all its arsenal to stop the renaming. Nigeria went to court in New York to block the naming of the street corner after the slain activist. A non-governmental organisation created and funded covertly by the Abacha regime, Congress of Nigerians Abroad (CNA), argued in court that the proposed naming was an 'unconstitutional attempt by the New York City to conduct foreign policy.'[16]

On 17 September 1997, the Council listened to the case for the naming. Former New York Mayor who is internationally respected for his leadership role in the anti-apartheid movement, David N. Dinkins, joined by Wole Soyinka, the Nigeria Nobel-prize winning playwright and activist – testified in favour of the re-naming. Nigeria's Ambassador to the UN and former External Affairs Minister, Professor Ibrahim Gambari, led the Nigerian officials who spoke against it. In 25 October 1997, at another hearing, the recently retired US Ambassador to Nigeria, Walter Carrington, who had openly identified with the pro-democracy activists while in Nigeria, and whose farewell party organised by the pro-democracy activists was disrupted by the agents of the Abacha regime, told the New York City Council that he was convinced that Nigeria's military dictators killed Kudirat Abiola.[17] On his part, exiled leader of the coalition of pro-democracy movements in Nigeria, the National Democratic Coalition (NADECO), the septuagenarian Anthony Enahoro, delivered a moving appeal to the Council: 'Mr. Chairman and Council Members, yours is the historic opportunity to accord enduring recognition to an inspiring

[14]Jumoke Ogunkeyede, Chairman, UCSN, personal discussions, New York, USA, 19 December 1999; Interviews: Jumoke Ogunkeyede, Ibadan, Nigeria, 4 July 2011; Remi Oyeyemi, former Director of Media Services, UCSN, Wilmington, Delaware, 27 April 2011.

[15]Bernard Stamler, 'The Fight over Kudirat Abiola Street', *New York Times*, 5 October 1997. http://www.nytimes.com/1997/10/05/nyregion/neighborhood-report-east-side-the-fight-over-kudirat-abiola-st.html (visited 12 February 2015).

[16]'Street Corner Named for Nigerian Dissident', *New York Times*, 25 January 1998. http://www.nytimes.com/1998/01/25/nyregion/metro-news-briefs-new-york-street-corner-named-for-nigerian-dissident.html (visited 12 February 2015).

[17]Duro Adeseko, 'Securitymen Stop Party for Carrington', *The Guardian*, [Lagos] 19 September 1997, p. 1; Bosun Odedina, 'Police Stop NADECO Party for Carrington', *Nigerian Tribune* [Ibadan], 19 September 1997, p. 1.

heroine, a victim of an awesome tragedy, a young African woman who dared to confront a brutal military regime and demand democracy and human rights for the Nigerian people – Kudirat Abiola.'[18]

After Carrington and Dinkin's testimony in support of the motion to name the corner after Abiola, the Council voted unanimously to do so. The vote was considered 'a major political defeat for the Nigerian military government.'[19] The new street sign was installed on 24 January 1998 at a ceremony where Kudirat Abiola's daughter, Hafsat Abiola, a Harvard-educated activist, unveiled the name plate. The campaign led by the Nigerian pro-democracy activists in the US, and which was regularly and prominently reported in Nigerian newspapers, constituted part of what AbdouMaliq Simone would describe as 'provisional, highly fluid, yet coordinated and collective actions'[20] that were generated in the quest for the humanisation of the Nigerian space, in ways that would lead to its continued existence and survival along the lines of bureaucratic rationality and democratic ethos. This was evident in the democratic process that produced the renaming of the NYC corner after the slain activist. It is a form of politics that employs naming as a way of shaming in the global public sphere (Fig. 14.1).

'Geographical names are apt to raise passions to a high level, and sometimes to the boiling point', argues Naftali Kadmon, adding that 'hardly anywhere is this more apparent than in the use of toponyms in a political environment or context.'[21] The naming was a crisis for Abacha's Nigeria. It was the most direct and the most significant symbolic 'diplomatic' slap on the face of Abacha's regime to be constructed in (or by, as the regime saw it) one of the most powerful nations in the world. The travel ban on top military officers and their family members did not attract comparable rancour from the Abacha regime. Even though the US federal government was not involved in the street-corner naming, the Abacha regime and the Congress of Nigerians Abroad, a pro-Abacha group, regarded it as a manifestation of the Clinton administration's 'foreign policy' towards Abacha's Nigeria. For the Abacha regime, the naming not only recognised, celebrated and 'consecrated' an assassinated arch-enemy, but also gave the 'dissidents' (the exiled pro-democracy activists) who were campaigning against it a visibility, audibility, credibility, and legitimacy that the regime had tried so cruelly to deny them. New York City was the last place where the regime would have wanted to be so humiliated. Through

[18]Murray Seeger, 'Nigeria Descends Deep into Disrepute: Rule by "Medieval Warlords" is Holding Back Largest Nation on the African Continent', *Baltimore Sun*, 21 December 1997. http://articles.baltimoresun.com/1997-12-21/news/1997355042_1_nigeria-abacha-walter-carrington/2 (visited 12 February 2015).

[19]'U.S. Ambassador "Convinced" Nigerian Regime Murdered Opponent – New York City Votes to Honor Slain Democracy Leader', *The Africa Fund*, 27 October 1997. http://www.africafocus.org/docs97/nig9711.php (visited 12 February 2015).

[20]AbdouMaliq Simone, *For the City Yet to Come: Changing African Lives in Four Cities* (Durham, NC: Duke University Press, 2004), p. 13.

[21]Naftali Kadmon, *Toponymy: The Lore, Laws and Language of Geographical Names* (New York: Vantage Press, 1997 [2000]), p. 79.

Fig. 14.1 Kudirat Abiola Corner, New York City (photo by Yele Sowore)

the naming of a corner in her honour and memory, Kudirat Abiola's physical extermination was transformed into a permanent symbolic presence in a global city.

14.3 The Toponymy of Tyranny: Louis Farrakhan Crescent in Lagos

The naming of the street corner in New York City seemed to have taught the Abacha regime a lesson in toponymy as foreign policy by other means. A humiliating toponymic 'coup' was successfully executed against the Nigerian serial coup-plotter in New York City on 24 January 1996 – to the acclaim of the free world. The regime's homicidal practices, its repressive apparatuses and its entrenched resistance to democracy, liberty and justice had just been loudly condemned in a global city. As a discursive, as well as material, non-amicable action in response to the conduct – or what is considered by one state to be the misconduct – of another state, street (re)naming can constitute a social and political attempt at the conduct of (mis)conduct. This is the way in which Nigerian activists and their local supporters in the US can be read to have approached the naming of Kudirat Corner, and this is also the way in which I read the reprisal by the Abacha regime in Lagos, Nigeria. Because the regime regarded the naming of the corner in New York City as injurious, the way in which it attempted to govern that otherwise ungovernable, but injurious, interference in the internal affairs of Nigeria by the New York City Council (and therefore, in the regime's estimation, the US Government) was to respond in like manner.

In constructing its response to the 'retortion', the Abacha regime decided to use the earlier visit of the anti-establishment African American activist, Louis Farrakhan, to Nigeria in February 1996, about four months before Kudirat was assassinated. During Farrakhan's visit, Abacha had urged him to assess Nigeria for himself and not rely on the western media reports about a country which, in *Washington Post*'s report of the Farrakhan-Abacha encounter, described as 'a nation shunned internationally for human rights abuses and lack of democracy.'[22] The influential newspaper added that General Abacha's guest was 'reviled by many for his anti-white and anti-Semitic views.'[23] Farrakhan's 27-day 'World Friendly Tour' of African and Middle Eastern countries included Libya, Iran, Iraq, Sudan and Syria, all countries described by the US government as 'sponsors of terrorism.'[24] During his six-day state-sponsored visit to Nigeria, Farrakhan decried western countries for their criticism of Abacha's execution of the writer, Ken Saro-Wiwa and the other Ogoni minority rights activists.[25]

Two weeks after the Kudirat Abiola Corner's sign was raised in New York, the Abacha regime secretly ordered the military appointed Eti-Osa Local Government Council in Lagos State – a state which was the hotbed of opposition politics – to rename Eleke Crescent in Victoria Island, Lagos, after the Leader of the Nation of Islam and virulent critic of the US government, Farrakhan (Fig. 14.2).[26] The foreign embassies located on Eleke Crescent included those of the United States, Britain, Canada, The Netherlands, and India. It was the regime's own way of shaming the US government in return. In a brief ceremony, which newspapers reported as 'appearing [...] to be a retaliation', the chairman of the council, Yahaya Adeniyi Dosunmu, described Farrakhan as 'a friend indeed by ensuring the preservation of the unity of this country, by being an impartial observer as regards the political impasse witnessed in the country a few years ago.'[27] It was obviously a script written for Dosunmu by the military regime in Abuja. To give the false impression that the renaming was not a retaliatory measure, Dosunmu said that the process of renaming the crescent had started a long time ago. However, he testified to the strategic importance of the street when he stated that 'This street (crescent) is very important and strategic to us. It houses a lot of embassies and the man we are honouring is an international man.'[28]

[22] 'Farrakhan Visiting Nigerian Leadership', *The Washington Post*, 3 February 1996.

[23] 'Farrakhan Visiting Nigerian Leadership', *The Washington Post*, 3 February 1996.

[24] See: 'Travels with Tyrants: Minister Louis Farrakhan's Anti-American World Tour' http://www.adl.org/travels_with_tyrants/travels_with_tyrants_old.asp (visited 12 February 2015).

[25] 'Travels with Tyrants.'

[26] Dozie Okebalama (et al.), 'Kudirat Corner: Nigeria Retaliates, Names Eleke Crescent after Farrakhan', *Punch* [Lagos], 6 February 1998, p. 1.

[27] Bosun Odedina, 'Kudirat Abiola Corner: Nigeria Hits Back', *Nigerian Tribune*, 6 February 1998, pp. 1–2.

[28] Odedina, 'Kudirat Abiola Corner', p. 2.

THE PUNCH

FRIDAY, FEBRUARY 6, 1998

Clinton, Blair confer over Iraq

Kudirat Corner

Nigeria retaliates, names Eleke Crescent after Farrakhan

LOUIS FARRAKHAN CRESCENT

Diplomatic Corps, NADECO react

Fig. 14.2 Front page of *The Punch* newspaper, Lagos, 6 February 1998 (author's collection)

While Abacha's supporters praised the renaming, pro-democracy activists in Nigeria condemned it. The leader of the frontline pro-democracy movement, the National Democratic Coalition (NADECO), Chief Abraham Adesanya, described the renaming as 'thoughtless.' While assuring the world that the street would be renamed 'when a new government comes',[29] Adesanya added that if the Abacha regime thought it would 'fight the American government with Farrakhan' it was making a mistake. The Campaign for Democracy (CD), a key civil society organisation, agreed with Adesanya by also predicting that the 'joke' would be 'reversed once Nigeria returned to democratic order.'[30] The retaliatory street-renaming was described by the British Broadcasting Corporation (BBC) as 'a street-naming tug-of-war.'[31] This was a clear case of toponymy as war by other means. Like the naming of the corner in New York to honour 'the indefatigable visionary leader' (Kudirat Abiola), the renaming of Eleke Crescent in Lagos also attracted the attention of the world's media. For instance, in London *The Independent* described it as 'political street game.'[32]

[29]Odedina, 'Kudirat Abiola Corner', p. 2.

[30]'Groups Flay Farrakhan Crescent', *National Concord* [Lagos], 9 February 1998, p. 2.

[31]British Broadcasting Service, 'Nigeria Names Street after Farrakhan', 6 February 1998. http://news.bbc.co.uk/2/hi/africa/54135.stm (visited 12 February 2015).

[32] 'Political Street Game', *The Independent* [London], 7 February 1998. http://www.independent.co.uk/news/political-street-game-1143295.html (visited 12 February 2015).

Thus, in the glocal politics of who deserves a place of honour and memory in the urban landscape and official street-naming, it can be argued that the Louis Farrakhan Crescent in Lagos was named in relation to the Kudirat Abiola Corner in New York. The US Mission to Nigeria in Lagos was counterpoised to the Nigerian Mission to the United Nations in New York. Even though the US was a superpower, Abacha's Nigeria, by this act of 'deviance' of the US, was stating its notional equality to the US – in the comity of nations. Also, within its own territorial space, Abacha's Nigeria was 'affecting' US policy by naming a street after an anti-establishment activist, one who constantly denounced the US government. Thus, the regime was shaming the US, just as it assumed that the US government 'affected' its own foreign policy by shaming it through the naming of Kudirat Abiola Corner in Manhattan. Again, while the renaming of Kudirat Corner can be interpreted as further democratisation, liberalisation and globalisation of the New York street corner, this can be contrasted with the production of a racialised landscape that the naming of Farrakhan Crescent could be said to denote.[33] On both ends therefore, there were attempts at the conduct of (what is assumed to constitute) local and (inter-)national 'misconduct' – in the case of Nigeria, the treatment of Nigerians in general and opposition elements in particular; and in the US case, the treatment of African Americans, both historically and in the contemporary US.

However, throughout the short period (February 1997 to May 1999) when this street officially bore the name of Louis Farrakhan, the US Embassy never used the name, retaining instead the name of Eleke Crescent as its official address. This refusal by both the US and other embassies to use the address Farrakhan Crescent was also a way of contesting the power of the Abacha regime to name their location. As Maoz Azaryahu argues, avoiding the use of particular street names 'amount to civil disobedience' and can also constitute 'a profound act of resistance.'[34]

14.4 Toponymy of Democracy: Walter Carrington Crescent in Lagos

Many Nigerians insisted that what happened on 8 June 1998 was a 'miracle.' The sudden death, on this day, of the Nigerian dictator, General Sani Abacha, who had practically kept the nation in jail while stealing billions of dollars, produced celebrations across the country.[35] With his death, the country was able to breathe again. His successor, General Abdulsalami Abubakar, released several political

[33] Reuben Rose-Redwood, Derek Alderman, and Maoz Azaryahu, 'Geographies of Toponymic Inscription: New Directions in Critical Place-Name Studies', *Progress in Human Geography*, 34, 4, (2010), pp. 453–470.

[34] Maoz Azaryahu, 'The Power of Commemorative Street Names', *Environment and Planning D: Society and Space*, 14, 3 (1996), pp. 311–330 (p. 318).

[35] James Rupert, 'Nigerian Ruler Dies after Brutal Reign', *Washington Post*, 9 June 1998, p. A1.

prisoners and encouraged exiled pro-democracy activists to return home. He also announced a short programme of transition to civil rule which was successfully completed on 29 May 1999.

One of the manifestations of this stage of 'liminal transition in [the] political history'[36] of Nigeria and the proclamation of a new era, was the renaming of Farrakhan Crescent by the newly sworn-in Governor of Lagos State, Senator Bola Tinubu, who himself had recently returned from exile. Tinubu was involved in the mobilisation for the naming of Kudirat Abiola Corner in New York and was therefore familiar with the retaliatory renaming of Eleke Crescent. As soon as he took power in Lagos State in May 1999, he embarked on the naming and renaming of some streets to signpost not only the beginning of a new democratic era, but also to serve as a reminder of the recent history of democratic struggle against military fascism.

We can therefore understand the basis of the few streets that were renamed by the Tinubu administration in Lagos State, and the few statues constructed or reconstructed by the new democratic government in Nigeria's old capital city – as well as in Ibadan, the capital of the old Western Region.[37] Six days after he was sworn-in as the democratically-elected governor of Lagos State, Governor Tinubu renamed Louis Farrakhan Crescent after the former US Ambassador to Nigeria and friend of the pro-democracy activists, Walter Carrington, who he described as 'a supporter of good government' in Nigeria.[38] Tinubu later added that, 'We needed a voice for the international community. Ambassador Carrington was such a voice. He didn't let diplomatic courtesy stand in the way of pursuing human liberty as the representative of the world's most powerful nation.'[39]

The renaming of Farrakhan Crescent as Carrington Crescent thus was both 'a celebration of the triumph' of democratic forces in Nigeria – as symbolised by the former US envoy – and it was 'a mechanism for settling scores with the vanquished regime',[40] which had driven the new governor, as well as his political associates in the opposition movement, into exile for more than 4 years. It also constituted a form of erasure of a version of the past – as represented by Farrakhan, who, is this context, symbolised Abacha's vicious rule. Therefore, dropping Louis Farrakhan name was a form of de-commemoration.[41] It was a way of dumping him in the 'garbage can' of history reproduced by the new inheritors of power in Lagos State. Also, in a sense, renaming the street after Carrington was a way of democratising the urban space in

[36]Azaryahu, 'The Power of Commemorative Street Names', p. 321.

[37]See: Wale Adebanwi, 'The Cult of Awo: The Political Life of a Dead Leader', *Journal of Modern African Studies*, 46, 3 (2008), pp. 335–360.

[38]'Tinubu turns Farrakhan Road to Carrington Way', *Tribune on Saturday* [Ibadan], 5 June 1999, p. 10.

[39]Interview, Senator Bola Tinubu, 2 July 2011, Lagos, Nigeria.

[40]Azaryahu, 'The Power of Commemorative Street Names', p. 322.

[41]Maoz Azaryahu, 'German Reunification and the Politics of Street Names', *Political Geography*, 16, 6 (1997), pp. 479–493 (p. 481).

post-military Lagos. This time, as a confirmation of the legitimacy and popularity of the renaming, the US Consulate and all the other embassies on the street started using the new name.

14.5 Conclusion: Toponymic Practices, Democratic Politics and Glocal Relations

Toponymic practices can constitute 'political tool[s]'[42]; or, put another way, they can constitute political struggles dressed in cartographic clothes. Both hegemony and counter-hegemony, inclusion and exclusion, incorporation and expulsion, and cooperation and conflict do speak through space.[43] I have illustrated how street naming and the associated renaming can yield both progressive and regressive consequences for social actors, regimes, groups and nations.[44] This chapter points to the imbrications of street names in conflictual relations and social and political struggles at the local, national and international stages. The examples illustrate the imbrications of the global in the local and the local in the global, which produce certain forms of glocality. I have also attempted to point further towards the inter-textuality of street naming in the context of symbolic (and material) (inter-)national relations – or what Joshua Hagen describes as the 'scalar politics of toponymy.'[45] In this context, we can understand scale not only in terms of its physicality and hierarchy, but in terms of its relation, interaction, and interconnection with other scales.[46] Such approach to scalar relations helps us to connect Kudirat Abiola Corner in New York to Louis Farrakhan Crescent and Walter Carrington Crescent in Lagos, and to show how states, governments, democratic groups and others deploy scalar politics and practices in order to 'crystallize certain socio-spatial arrangements in consciousness and [...] further social, political or cultural aims.'[47] Toponymy can be thus approached as a specific form of (spatial) governmentality – which is conceived here as the conduct of (mis)conduct. By speaking with space through street naming, the Nigerian exiles and pro-democracy activists, the military regime and the succeeding democratic government all exhibited the spatiality of

---

[42]Kadmon, *Toponymy*, p. 79.

[43]Garth Andrew Myers, 'Naming and Placing the Other: Power and the Urban Landscape in Zanzibar', *Tijdschrift voor Economsche en Sociale Goegrafie*, 87, 3 (1996), pp. 237–246 (p. 244).

[44]Robin Kearns and Lawrence Berg, 'Towards Critical Toponimies', in Lawrence Berg and Jani Vuolteenaho (eds), *Critical Toponymies: The Contested Politics of Place Naming* (London: Ashgate, 2009), pp. 1–18 (p. 1).

[45]Joshua Hagen, 'Theorizing Scale in Critical Place-Name Studies', *ACME: An International E-Journal for Critical Geographies*, 10, 1 (2011), pp. 23–27 (p. 26).

[46]Hagen, 'Theorizing Scale', p. 24, following Richard Howitt, 'Scale as Relation: Musical Metaphors of Geographical Scale', *Area*, 30, 1 (1998), pp. 49–58.

[47]Adam Moore, 'Rethinking Scale as Geographical Category: From Analysis to Practice', *Progress in Human Geography*, 32, 2 (2008), pp. 203–225.

power relations. Street naming can be used to participate in geopolitics through the emblazoning of the interplay of national and supranational relations on local spatial referents, yet it does not operate in isolation.[48] In illuminating the complexities and dynamics that are enfolded in the practices of naming, this chapter has shown that street naming is connected to or enfolded in existing social relations, both internal and external. It is also a manifestation of discursive power relations, and even a form of retortion.

[48]Rose-Redwood, 'From Number to Name', p. 436.

Chapter 15
Afterword

Ambe Njoh and Liora Bigon

Abstract This short chapter concludes our team effort of elaborating on the grammar of toponymic expressions in urban Africa, historicise and contextualise their inherent characteristics, ambiguities and tensions, and exposing their capillary power. It calls for further works focusing on the global South, in order to enrich our toponymic insights and historiographies by closing the gap between the northern and the southern hemispheres.

Keywords Global South • Toponymic legacies • African urban spaces • Entangled histories • Toponymic varieties • Colonialism

Place names and (colonial) urban legacies are obviously universal phenomena. Thus, the main question to be asked regarding this unprecedented collection relates to the relevancy of the African continent to toponymic studies in general. At this point, we do not think that investigating these phenomena is advantageous because they are located in 'Africa', but rather because the variegated African experience shares common-cum-particular aspects. Amongst them are historical (pre-colonial, colonial, post-colonial), historiographic (research tendencies and traditions, research prejudice), methodological (enriching experiences of creative usage in multidisciplinary sources and interconnected evidence), and actual practices (mobile and ambiguous toponyms; top-down and bottom-up, generic and specific, modes of naming; a strong interface between the ancient and the modern, the autochthonous and the occidental – though often through the 'distorted' colonial channel). These common-cum-particular aspects in the urban toponymic experience of Africa challenges or counter-balances the mainstream of critical place names studies. The latter studies are preoccupied with Western and Eastern Europe, North America and 'nationalism', and are conducted with somewhat monolithic

A. Njoh (✉)
Department of Geography, University of South Florida, Tampa, FL, USA
e-mail: njoh@usf.edu

L. Bigon
The Institute of Western Cultures, The Hebrew University of Jerusalem, Jerusalem, Israel

Holon Institute of Technology, Holon, Israel
e-mail: liorin@hotmail.com

L. Bigon (eds.), *Place Names in Africa*, DOI 10.1007/978-3-319-32485-2_15

methodology (see our Introduction). But, at the same time, these 'African' aspects also testify to the validity of another key idea that underlines this collection and transcends beyond its seemingly 'area-studies-ness', that is, the idea of entangled histories.

Our team effort to elaborate on the grammar of toponymic expressions in urban Africa, historicises and contextualises its inherent characteristics, ambiguities and tensions and expose its capillary power – can not be reduced into certain geographical demarcations. As remarked by Sanjay Subrahmanyam, though with respect to the early modern history of mainland South East Asia: "Given the fragmentary nature of access to knowledge, each of us is more or less condemned in greater or lesser measure to Area Studies." Consequently, according to him, not only a comparative approach should be pursued, but rather the seeking out of the,

> at times fragile threads that connected the globe, even as the globe come to be defined as such. This is not to deny voice to those who were somehow 'fixed' by physical, social and cultural coordinates, who inhabited 'localities' in the early modern period and nothing else, and whom we might seek out with our intrepid analytical machetes. But if we ever get to 'them' by means other than archaeology, the chances are that it is because they are already plugged into some network, some process of circulation.[1]

While these inspiring words ring almost true for the pre-colonial period, they become obvious under the mechanism of modern colonial urbanism. These urban spaces in Africa constituted primary *lieux de colonisation*, established by the leading European forces of the mid-nineteenth century as a means of constructing and channeling the global economy. As such, African urban spaces bridged upon complex processes of material and conceptual flows – also in terms of toponymy as exemplified in each contribution in this volume – which essentially stretched much beyond their immediate site-relatedness. Connecting together local, regional, national, transnational and international frontiers, the complexity becomes only further elaborated in the post-colonial period, when topnymic legacies continue into the present. Reflecting *lieux de mémoire*, Africanisation, Arabisation, re-colonisation, (inter-)national retortion and transformation policies or simply a laissez-faire, post-colonial toponymy converses with new challenges presented by modern urban management, changing political regimes and international relations. Yet, the 'entangled' is not rooted solely in the question of geographies, scales, time spans and relevant histories and legacies that influence toponymic practices. It is also related, as elaborated in this collection, to the very toponymic features and the nature of spatial terminologies in question. For instance, the interplay between the oral and the inscribed, the fixed and the shifting, the uncanny and the remote, the generic and the specific, the exogenous and the indigenous, the mundane and the ceremonious, the commemorative and the environmental, the hegemonic and the counter-hegemonic.

[1]Sanjay Subrahmanyam, 'Connected Histories: Notes towards a Reconfiguration of Early Modern Eurasia', *Modern Asian Studies*, 31, 3 (1997), pp. 735–762 (pp. 761–762).

This relational and inclusive approach opens another forum for entanglements: of intercrossing in between these various aspects as they run across the present anthology. But more than using the 'entangled' here in order to serve as an analytical toolbox for systematic analysis,[2] we rather hope to encourage further discourse in relation to other regions and experiences in and far beyond Africa – this could be also associative and abstract. Aware of the vivid potential of place names studies in general and as a burgeoning field of multidisciplinary research in particular, this collection invites further works focusing on the global South. Enormously expand the subject matter, these future works would enrich our toponymic insights and historiographies by closing the gap between the northern and the southern hemispheres. But, as concluded by Steven Feierman in an essay entitled 'African histories and the dissolution of world history': "Historians have no choice but to open up world history to African history, but having done so, they find that the problems have just begun."[3]

[2]As suggested, regarding the notion *histoire croisée*, in: Michael Werner and Bénédicte Zimmermann, 'Beyond Comparison: *Histoire Croisée* and the Challenge of Reflexivity', *History and Theory*, 45, 1 (2006), pp. 30–50.

[3]Steven Feierman, 'African Histories and the Dissolution of World History', in: Robert Bates, V. Y. Mudimbe and Jean O'Barr (eds), *Africa and the Disciplines: The Contributions of Research in Africa to the Social Sciences and Humanities* (Chicago: The University of Chicago Press, 1993), pp. 167–212 (pp. 198–199).

Index

L. Bigon (eds.), *Place Names in Africa*, DOI 10.1007/978-3-319-32485-2

The manufacturer's authorised representative in the EU is Springer Nature Customer Service Centre GmbH, Europaplatz 3, 69115 Heidelberg, Germany. If you have any concerns regarding our products, please contact ProductSafety@springernature.com

Printed and bound by CPI Group (UK) Ltd, Croydon, CR0 4YY
30/09/2026
02225571-0004